CALIFORNIA'S
SALMON AND STEELHEAD

CALIFORNIA'S SALMON AND STEELHEAD

The Struggle to Restore an Imperiled Resource

EDITED BY ALAN LUFKIN

UNIVERSITY OF CALIFORNIA PRESS
BERKELEY   LOS ANGELES   OXFORD

Royalties from the sale of this book are dedicated
to education programs for the restoration of California's
salmon and steelhead stocks.

University of California Press
Berkeley and Los Angeles, California

University of California Press, Ltd.
Oxford, England

Library of Congress Cataloging-in-Publication Data

California's salmon and steelhead : the struggle to restore an
    imperiled resource / edited by Alan Lufkin.
        p.   cm.
    Includes bibliographical references (p.) and index.
    ISBN 0-520-07028-3 (cloth)
    1. Salmon-fisheries—California.   2. Steelhead (Fish)—California.
I. Lufkin, Alan.
SH348.C35   1991
333.95'616'09794—dc20

                                                    90-12747
                                                      CIP

9 8 7 6 5 4 3 2 1

Contents

Foreword

Anthony Netboy

There has long been a need for a comprehensive and readable book on the unhappy fate of California's salmon and steelhead trout. Alan Lufkin and his associates have provided it, and we should be most grateful to them.

A lecture I gave many years ago in England is entitled "Man and the Salmon: A Problem of Coexistence." The story of the salmon in California is typical of what the title implies. The salmons were probably distinct species over a million years ago, when the advanced anthropoid apes were on their way to becoming humanoids. Man and the salmon had a harmonious relationship until recent centuries. Now the salmon are being increasingly harassed by man the world over and coexistence is becoming more difficult, especially in industrial countries. And, I may add, nowhere is the existence of the anadromous fishes more difficult than in California.

There is a difference in the way these fishes were treated by the aborigines and the white man. Coastal Indians and those living in the Sacramento/San Joaquin valleys treasured the abundant salmon that lived in the streams almost the year round. "No stream was too small to host populations of these hardy fishes, and the supply seemed endless," says Lufkin. The Indians were keenly aware of the importance of salmon for their survival. They did not take the bounty for granted. With their mystic sense they developed rituals and myths that they believed would assure abundant runs. Moreover, they became conservationists and did not waste them.

This became clear to me from an interview with a Tlingit Indian in Alaska, descendant of chiefs. He said, "The Great Creator, my father told me, sees everything. The undying Creator created the fish for the benefit of human beings, but we must not take them except for food. In Sitka," he added, "they used to destroy three scows of salmon at a time because the canneries could not handle them. We were taught it was a sin to kill off the seed stock, but the white man killed the seed stock and depleted the rivers."

The story unfolded in this book is part of what the California ecologist Raymond Dasmann calls "the destruction of California," the title of his book. The fate of California's salmon mirrors the state's use of its environment and natural resources, especially water, which is vital to the existence of both fish and men.

In the past century a land of infinitely varied landscapes, endowed with an abundance of fertile soils, forests and grasslands, mountains and deserts, and countless rivers, was invaded by millions of people from the four corners of America, seeking a better life in a milder climate. In the process rivers and watersheds were turned topsy-turvy; farmland was bulldozed to make way for human habitations. Forests containing trees hundreds of years old were reduced to lumber and other forest products; desert lands were trampled to dust; foothills and lowlands were occupied by housing developments; rivers were dammed to generate power, and in the process prevented the migration of anadromous fish to their spawning grounds. Tremendous amounts of water were diverted to irrigate semidesert land to grow needed crops and also to grow cotton that became a drug on the market.

Most of California's salmon and steelhead were, so to speak, evicted from their native habitats, and the runs declined or disappeared. Only a fraction of the original cornucopia remains.

This book documents the story and pinpoints the way Californians have mistreated and exterminated most of the state's salmon and steelhead runs. The engineers who ran the Bureau of Reclamation that built the great Central Valley water projects and others had little interest in saving the fishes. "Bureau policies made fisheries expendable," says Lufkin. "While national emergency restrictions could partly explain the bureau's earlier neglect of fishery values, that excuse was invalid. Had the bureau genuinely acknowledged fish and wildlife values, fish protection planning could have begun with preliminary engineering studies and been realistically paced throughout the planning process. That did not happen."

What stopped some projects that would have been harmful to the fisheries was due largely to citizens' agitation resulting in action taken by the state legislature to establish citizen advisory committees. The California Department of Fish and Game had little power and less money. It could offer only technical help in planning mitigation facilities.

"The Bureau of Reclamation," says Lufkin, "is widely blamed for subsequent fishery declines traceable to the Central Valley Project, and with reason. The agency gave lip service to fishery conservation causes, but its action belied its words." The Shasta Salvage Plan failed, and the San Joaquin fishery died. Trinity River salmonid stocks were decimated. The bureau's primary policy was to provide water for agricultural irrigation, and this aim precluded improvement of fisheries. In brief, the value of fisheries was downplayed in favor of benefiting the agribusiness interests.

I cannot help contrasting the dismal failure in California to protect the fisheries with the success attained in the Pacific Northwest to force the dam builders, mainly the powerful and arrogant Corps of Engineers, to build fish ladders, bypasses, and other facilities to protect the salmon. This victory was not attained without a struggle. It is hard to believe that the original plans to build Bonneville Dam, the corps' first project on the Columbia, excluded facilities for protecting the fish. When fishery people protested, the chief of the corps reportedly said, "We don't intend to play nursemaid to the fish." Had this policy prevailed, the entire cornucopia of salmon and steelhead in the vast Columbia River watershed would have been doomed. Strong public opposition, however, forced the corps to add fish-saving facilities at each of the many dams built on the Columbia. On the Snake River, the largest tributary of the Columbia, no workable facilities were provided at the Idaho Power Company's three dams, and the salmon and steelhead were exterminated.

Having visited most of the countries in the Northern Hemisphere that have seagoing salmon, and written four books about them, I have concluded that man and the salmon are indeed on a collision course. Salmon are the world's most harassed fish, the title of one of my books. The story of their fate in every industrial country is the same: downbeat. In other words, man and the salmon cannot live harmoniously together in such countries. Where the citizens have enough interest to protect the fisheries at dams and other impoundments there is a good chance most of them will be saved, as on the Columbia River. Where the people are lax, or dominated by water and similar interests, the fishes will largely go down the drain, as in California.

Acknowledgments

This volume represents the efforts of dozens of people who helped it evolve from inspiration to bound reality. I share credit with them for its merits.

As the book took shape, I was reminded repeatedly that most people are naturally inclined to be cooperative. Fishery professionals, librarians, state water and forestry officials, secretaries in a variety of offices, members of various salmon and steelhead restoration groups—the list could go on and on—have been so helpful that I regret being unable to name them all here. Their kind words and helpful acts have been deeply appreciated by this no doubt unremembered fellow who appeared, sometimes unexpectedly, in their offices or introduced himself and his concerns by telephone. Throughout the project, staff of the California Department of Fish and Game and the U.S. Fish and Wildlife Service were particularly helpful.

I feel special gratitude toward those who spent valuable hours writing chapters for this volume: Scott Downie, Joel Hedgpeth, Pat Higgins, Bill Hooper, Dick Hubbard, Bill Kier, Bill Matson, Paul McHugh, Phil Meyer, Ronnie Pierce, Kim Price, Nancy Reichard, Dave Vogel, Cindy Williams, Jack Williams, and Bob Ziemer. Many of these authors also found time to review drafts of various chapters and make valuable suggestions. They did this not for money but because of their concern for California's salmon and steelhead.

I also offer most sincere thanks to the authors of several published pieces included here. These writers not only consented to the use of their material, but in several cases they worked with me to make minor changes in their contributions. Several added interesting follow-up observations. Included in this group are Stan Barnes, Bill Davoren, Dick Hallock, Ken Hashagen, Joel Hedgpeth, Eric Hoffman, Bill Poole, Felix Smith, Bill Sweeney, and George Warner. Through the courtesy of Jim and Judy Tarbell, publishers of *Ridge Review*, I was able to obtain the report of an interview with former Senator Peter Behr and Mary-Jo DelVecchio Good's essay on women and fishing on the North Coast. (Mary-Jo's cheerful willingness to reshape her chapter to focus on the salmon theme of the book was also most gratifying.)

Of inestimable value, too, were the contributions of the generous people—including a number of those named above—who helped in other ways: generously sharing their expertise, providing leads for further information, scrutinizing and commenting on various chapters, suggesting content, setting up field trips, and offering photographs. Among these individuals: Bill Bakke, Steve Canright, Marie De Santis, Katherine Domeny, Gary Flosi, Gene Forbes, Jack Fraser, Carl Harral, Herb Joseph, Karl Kortum, Mel Kreb, Howard Leach, Mark Lufkin, Paul Lufkin, Richard May, Wilmer Morse, David Muraki, Harold Olson, Marie Olson, Dick Pool, and Joey S. Wong.

I wish to thank Barbara Grant Lufkin, my wife, for her active, unfailing support of the project.

From my initial telephone contact through the sometimes arduous but always rewarding review process, the University of California Press provided warm support and expert editorial help in crafting the final manuscript into book form. Particularly helpful here were sponsoring editor Ernest Callenbach, copy editor Don Yoder, and project editor Mark Jacobs. University of California professors Don C. Erman and Peter Moyle, of the Berkeley and Davis campuses respectively, read the draft manuscript and offered excellent suggestions for improvement.

Anthony Netboy, a professor of English recognized worldwide for his many books and articles on salmon and natural resource management, supported the idea for the book from its inception. His continuing interest has been a source of great satisfaction.

I am especially indebted to fellow members of the California Advisory Committee on Salmon and Steelhead Trout, who welcomed the notion of a book on restoration of their favorite species and trusted me to put it together in my own way. Bill Kier, consultant to the Advisory Committee, was particularly generous with his time, providing in-depth answers to questions and helping me polish certain parts of the manuscript.

A.L.

Preface

Salmon and steelhead are premier California fishery resources that have been seriously harmed by man's shortsighted, exploitive approach to natural resource management. Today, however, because of these salmonids' commercial and sportfishing values, the fact that their well-being is an indicator of the health of many other species, and the special mystique of these beautiful fishes, diverse individuals and groups are joining together to help restore populations to viable levels.

The effectiveness of these efforts cannot yet be measured by dramatically increased total numbers of fish produced, although success stories, such as reports of record commercial catches, improved runs on the Klamath, and markedly improved sport catches off the Golden Gate, are increasingly heard. A promise of measurable success may be seen in the grubby, decidedly undramatic work of young people restoring damaged stream habitat; in professionals and academicians studying, experimenting, and arguing; in commercial, Indian, and sport fishermen joining together as fish activists; and in public leaders and staffs supporting a cause they see as important. These people share a simple goal: the numbers and quality of salmon and steelhead in California *will* increase.

Salmon and steelhead appeal to us all. The purpose of this book is to provide information about these salmonids for the general reader, information that will promote understanding and appreciation of the satisfactions, frustrations, and progress of efforts to save these valuable California resources.

The need for this collection of writings became apparent when the editor was serving on two public advisory committees: the Upper Sacramento River Salmon and Steelhead Advisory Committee and the California Advisory Committee on Salmon and Steelhead Trout. As a member of such committees, one soon collects an impressive volume—one member characterized it as "about a cubic yard"—of reading materials. These range from special scientific reports of the 1940s to environmental impact statements of the 1980s and include a variety of materials covering nearly every year of that period.

From perusal of such documents, two distinct insights emerge: a growing respect for salmon and steelhead (and the people who work to conserve their stocks) and the realization that no one person can be expert in all aspects of California salmonids. The subject is too vast, impinges on a hundred state economies, cuts across disparate cultures and life-styles, and requires the expertise of dozens of scientific fields. Perhaps most of all, because fish need water, salmonid conservation is always a contentious player in that long-running drama called California water politics. Even should one master the fundamental facts, keeping up with the underlying political-economic currents would be nearly impossible. To reconcile these constraints with the conviction that an aroused public should first be an informed public, the subject is presented by means of representative material. Hence the book is a collection of essays, excerpts, published articles, and speeches, selected to contribute uniquely toward an attainable goal: appreciation and understanding of one of California's most precious resources, the salmon and steelhead that live in her streams and ocean.

Another insight, more discouraging, also becomes apparent. One begins to comprehend the enormous power—economic, political, and social—wielded by forces that see salmonid restoration efforts as a simple choice between people and fish. The contributors to this book sometimes express despair over this viewpoint, which they consider irresponsibly inaccurate and dangerously simplistic. The philosophical question is much broader: Are Californians willing to make the choices necessary to assure a healthy environment for all living things? A retired Department of Fish and Game official summed up the problem this way: "Those responsible for fish have fought the good fight but have lost. All the restoration efforts and advisory committees are Band-Aids until the basic philosophy changes—and until it does, we stand to lose it all."

This collection will have served its purpose if it helps readers develop a new understanding of the choices to be made and influences prevailing philosophy so we will not lose it all.

Organization of the Book

The works of twenty-eight authors are represented here. They are arranged in four sets of perspectives: historical, current, legal-political, and local restoration. In organizing the diverse material in readable form, several problems arose that should be understood here at the outset. Because each author, although presenting a subtopic from his or her unique perspective, is treating the same central subject, some overlapping of material is inevitable. Problems related to fish hatcheries or the Central Valley, for example, are mentioned by several contributors. Editing reduced some overlaps, but to retain the integrity of each contribution, others had to be kept. Similarly, none of the book's four major divisions can be totally discrete from the others.

A final caveat: The book is intended to be neither a scientific treatise on salmon and steelhead (although many of the authors are scientists) nor an encyclopedic coverage of the subject. The materials range from scholarly discussions to passionate exhortations to get something done. They are intended to provide a rich, diversified selection of readings in which everyone can find something to like—and possibly to debate.

Outline of Salmon Biology

Although much salmon biology is revealed contextually in the various selections, some basic information may be helpful at the start. All six species of Pacific salmons have been reported in California waters, but only three common species are of concern here: chinook (*Oncorhynchus tshawytscha*), coho (*O. kisutch*), and steelhead (*O. mykiss;* formerly *Salmo gairdneri*). Other species—pink (*Oncorhynchus gorbuscha*), chum (*O. keta*), and sockeye (*O. nerka*)—are excluded because they rarely stray as far south as California. The genus name, *Oncorhynchus*, refers to a common characteristic of all these salmons: hooked nose, particularly apparent in males during spawning migrations.

Salmon are normally anadromous; that is, they are hatched in freshwater streams, mature in the ocean, and return—commonly to home streams at age three or four—to produce a new generation. Some chinooks and cohos, usually precocious males, return from the ocean at age two. Spawning typically takes place in cool, fast-moving water near riffles that keep eggs oxygenated. Streams with silt-free rocky/gravel substrates are essential for natural reproduction. With minor exceptions, all chinook and coho salmon, and most steelhead, varying from river to river, die soon after spawning.

Chinooks are both the most numerous and the largest California salmon. They are also typically "big river" fish, generally avoiding smaller coastal streams. Within a single river system there may be several distinct spawning runs of chinooks: fall, late fall, winter, and spring. Most California chinooks are fall spawners. Chinooks typically migrate to the ocean a few weeks after emergence from the gravel, while less than four inches long.

Coho salmon spend a year or more in fresh water before migrating to the ocean in the smolt phase. Since they must "summer over" in native streams, it is important that water temperatures not rise above seventy degrees. Therefore, cohos thrive best in cool coastal streams. They mature typically at three years of age and migrate upstream for spawning in fall and winter months.

Steelhead, until very recently, were considered a "cousin" of salmon. They are anadromous. They look like salmon. They spawn like salmon. Like cohos, they thrive in cooler coastal streams. But they exhibit greater variations in individual behavior than do salmon. For example, some steelhead are able to survive several spawning migrations, a trait of Atlantic, not Pacific, salmon. Some steelhead remain in fresh water for three or more seasons before migrating seaward, and a few steelhead offspring mature without ever migrating to sea—becoming, in effect, resident rainbow trout. Steelhead up to nine years old are known.

Spawning migrations of steelhead extend over longer periods than other salmon, sometimes ranging from May until the following spring. Because upstream migrations merge and the numbers wax and wane, almost never stopping completely, it is unrealistic to divide the runs into distinct seasonal events. This remarkable genetically endowed variability underscores the difficulties of attempting to reproduce viable runs of these fishes by artificial means.

A Note on Nomenclature

Formerly classified by the American Fisheries Society as *Salmo gairdneri,* suggesting their close relationship to Atlantic salmon (*S. salar*) and brown trout (*S. trutta*), steelhead are now, after exhaustive laboratory studies and review by professional fisheries, ichthyologist, and herpetologist societies, considered more closely related to Pacific salmon. Steelhead now share the genus *Oncorhynchus* with all other Pacific salmons. The new species name, *mykiss,* is the original Kamchatkan word for rainbow trout.

Most of the material in this book was written before the name change. Since "steelhead" remains in common use and the fish is still recognized as a sea-run rainbow trout, no changes in this nontechnical volume were deemed necessary. But the fact should be noted: a steelhead is, taxonomically, a Pacific salmon.

Another note relative to names: each species of Pacific salmons has several common names. The scientific community prefers the terms "chinook" and "coho" for the most common California salmon. The California Department of Fish and Game has traditionally used "king" and "silver," respectively, in references to these fishes, but a trend is growing to adopt the common names preferred by scientists. Therefore, "chinook" and "coho" are used most commonly throughout this volume. Two authors, Bill Kier and Bill Matson, strongly prefer the "king" and "silver" nomenclature because these terms are better known to consumers and politicians, the people who will determine the fate of these fishes. Their chapters use the traditional terms.

A.L.

Map 1 Rivers of California

Coastal and S.F. Bay Drainage			*Central Valley (ultimately S.F. Bay) Drainage*
1. Smith	10. Mattole	20. San Lorenzo	A. McCloud
2. Shasta	11. Noyo	21. Pajaro	B. Pit
3. Klamath	12. Big	22. Carmel	C. Sacramento
4. Salmon	13. Albion		D. Feather
5. Trinity	14. Navarro		E. Yuba
6. Little	15. Garcia		F. American
7. Mad	16. Russian		G. Mokelumne
8. Eel	17. Napa		H. San Joaquin
9. Van Duzen	18. Petaluma		I. Merced
	19. Guadalupe		

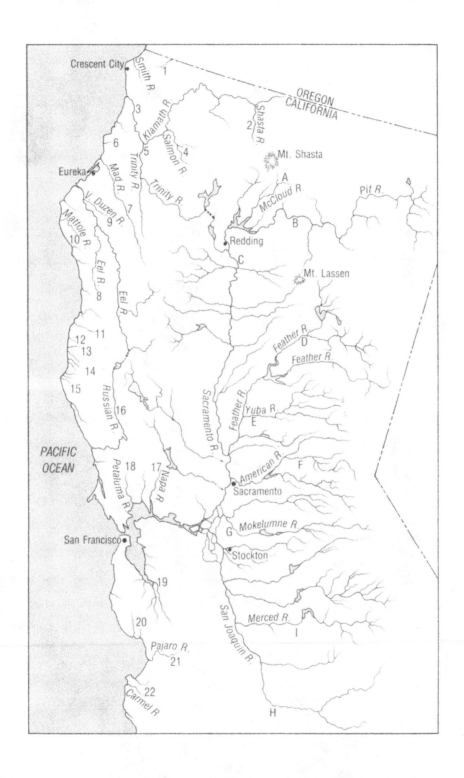

Crescent City

OREGON
CALIFORNIA

Smith R.

1

3

Klamath R.

Shasta R.

2

Mt. Shasta

5

Salmon R.

Trinity R.

4

6

Eureka

Mad R.

Trinity R.

A

McCloud R.

Pit R.

V. Duzen R.

7

B

Mattole R.

9

10

Eel R.

8

Eel R.

Redding

C

Mt. Lassen

11

12

13

Feather R.

D

14

Feather R.

15

Russian R.

16

Sacramento R.

Feather R.

Yuba R.

E

American R.

F

PACIFIC
OCEAN

Petaluma R.

18

17

Napa R.

Sacramento

Mokelumne R.

G

San Francisco

Stockton

19

San Joaquin R.

Merced R.

I

20

Pajaro R.

21

Carmel R.

22

H

Part One

Historical Perspectives

Part One consists of five chapters that focus on historical events, from earliest times to most recent, that have determined the current status of California's salmonid resources and suggest likely future developments. It introduces many of the topics that are later dealt with in greater detail.

Chapter 1 is a sketch of major historical developments affecting the fisheries from the early 1800s through 1989. The picture, most simply, is one of initial abundance followed by essentially unbridled harvesting of stocks and destruction of habitat as European immigrants crowded into California. Eventually, as near collapse of the fishery led tortuously toward development of protective legislation, California's commitment to statewide water development, principally to benefit agricultural interests, dealt fisheries a further blow. Competing statewide demands for water constitute the major problem today for those who would restore fishery resources. The historical sketch becomes detailed as it examines the greatly increased interest in salmon and steelhead resources of recent decades. In this review of landmark environmental legislation, both federal and state, we see how concerns for the subject are becoming central issues in California's precedent-setting environmental movement.

Chapter 2 discusses early Indian fishery problems on the Klamath River. Here Ronnie Pierce, a marine biologist and Indian historian, traces the changes that occurred in Native American life-styles as the invasion by non-Indians into their territory—in which there were only two directions, upriver and downriver—destroyed tribal structures. One may see startlingly how a benevolent but bumbling federal administration attempted to convert native fisher tribes to an agricultural life-style, how a sometimes venal Congress concealed its refusal to ratify Indian treaties, and how questions of ownership and control of tribal lands became mired in legal issues that only now are being effectively resolved. Forced to adapt to new ways, Indians became fishermen and plant workers for non-Indian canneries, but because their gillnet fishing was declared to be the cause of declines in runs of Klamath River salmon, that canning operation was closed by law in 1933. Problems springing from this painful history have persisted over the years and continue to be sore points in negotiations between Indian and non-Indian fishers.

In Chapter 3, North Coast fish restorationist Scott Downie discusses salmon and steelhead in a vignette about early settler families on the South Fork of the Eel River. To them, as with the Indians, the fish were both a source of food and an attractive diversion from rigors of life in the sometimes harsh unsettled lands. By the early twentieth century, it was becoming apparent that the influx of new settlers was introducing a way of life, symbolized by the artificial flies they tied, that portended unwelcome change. In this selection, Downie also tells much about the biology of salmon and steelhead, suggesting how these species evolved in the special conditions of northern coastal California, which also produced dense mixed conifer and hardwood forests, in addition to the tall redwood groves found along the rivers.

In Chapter 4, an excerpt from his essay first published in 1944, Joel W. Hedgpeth views the earliest efforts of government in the 1870s to develop artificial propagation of salmon. This study, a classic of salmon lore, also acquaints the reader with the salmon fishery of a Wintu Indian village on the McCloud River, a northern tributary of the Sacramento. It describes how a retired Unitarian minister and avid chess player, Livingston Stone, fascinated with the new science of "fish culture," attempted unsuccessfully to help restore salmon runs in Atlantic Coast rivers by introducing chinook salmon from Pacific Coast stocks. The site of the village and the hatchery is now hundreds of feet below the surface of Shasta Reservoir.

Chapter 5, the final historical chapter, is George Warner's poignant account of how the San Joaquin River spring chinook run was made extinct in the late 1940s as the water of that river was shunted south to Kern County as part of the Central Valley Project. The reader shares the hopes and frustrations of weary Division of Fish and Game crews as they rig war-surplus equipment to save a doomed species while poachers lurk nearby, devising ways to capture fish when crews are not looking. Although local sportsmen-activists won temporary fish protection flows from harassed Bureau of Reclamation functionaries, the combination of federal power, state acquiescence, and the water lobby ultimately extinguished the San Joaquin spring-run chinooks and essentially destroyed the remaining salmon fishery on that stream.

Joel Hedgpeth and George Warner have added intriguing epilogues to their stories.

Chapter One

Historical Highlights
Alan Lufkin

Salmon and steelhead have been significant anadromous fishery resources throughout California's history. The Native American people, totaling about three hundred thirty thousand individuals before the Europeans' arrival, depended heavily upon these fishes for subsistence, ceremonial, and trade purposes.

Early Indian Fishery

Indians of the Central Coast and in the Central Valley fished for salmon and steelhead found almost year-round in coastal streams and the Sacramento/San Joaquin rivers system. North Coast tribes, descendants of southerly migrating fishers, harvested these resources principally in late summer and fall months from the Eel, Klamath, Smith, and Trinity rivers and hundreds of lesser streams. It has been estimated that California's aboriginal peoples consumed or bartered about five million pounds of salmon per year. Salmon stocks functioned as a bank from which withdrawals could be made as needed: when other foods were in short supply, more salmon were taken. Tribal conservation practices, such as removing fish dams when adequate numbers of fish were taken, ensured that stocks would remain plentiful. No stream was too small to host populations of these hardy fishes; the commonly owned supply was endless. Methods of capture included several kinds of nets, spears, and a variety of wooden traps and weirs.

As highly valued resources, salmonid populations were the subject of Indian rituals and myths associated with the need to assure abundant runs throughout stream systems. The story of Oregos is one example. This helpful spirit in ancient times became a rock above the mouth of the Klamath River, where she told salmon about propitious weather conditions and warned them of dangers they must overcome. Today, when one thrills at the sight of salmon wildly pursuing anchovies below that rock, sometimes charging after individual fish for great distances along the surface, it is easy to dismiss the dictum that salmon stop eating when they enter the river. At that moment one may readily imagine that Oregos is telling the salmon to feed well on this, their last meal, as they prepare for their long and only upstream journey.

Immigrants Take Control

The period of non-Indian immigration and settlement during the nineteenth and early twentieth centuries was marked by a complex tangle of fishery-related matters. The subject is best presented in three overlapping perspectives: early development of the commercial salmon fisheries, effects of the gold rush and hydraulic mining, and early management efforts.

Commercial Beginnings

Early explorers and settlers of the westernmost frontiers were impressed with the quality and numbers of salmon and steelhead. Fremont ate "delicious salmon" at Sacramento. Bidwell and Sutter fished commercially, the latter employing Indian fishermen during the early 1840s. As part of the unexploited wealth of natural resources that drew immigrants to California, fish and wildlife initially belonged to no one and to everyone. Since the common resources seemed endless, harvest restrictions were unnecessary. Laissez-faire government encouraged exploitation of this cornucopia of resources for private gain and "public good."

In the absence of legal restraints, enterprising pioneers could fish almost as they pleased. In 1852 a state law enjoined "all good citizens and officers of justice" to "remove, destroy, and break down" any obstruction to salmon migration, except those erected by Indians. Other fishery laws soon followed, but such regulations were not enforced. Efforts to protect the Indians and their fisheries were farcical.

During succeeding decades large-scale commercial fishing was developed. Informal but sternly enforced agreements within certain ethnic groups, particularly Italians and Greeks fishing the Sacramento River, became the fisherman's law. Arthur F. McEvoy, in *The Fisherman's Problem,* details how places to fish, kinds of fish, and harvest limits were controlled by fishing communities. Exclusions were also imposed: Chinese, for example, were not permitted to fish for salmon. In this ethos, an unplanned, rudimentary system of conservation emerged; maintenance of the fisher community way of life was an important consideration. Both the economic well-being of fishermen and the biological well-being of fish stocks benefited. Today, fishermen's leadership in salmon restoration endeavors, and festive events such as Bodega Bay's annual salmon festival and Fort Bragg's Fourth of July salmon barbecue, remind us that modern California's sophisticated salmon fisher communities still value this tradition highly.

Another segment of the non-Indian fisher group, however, felt no such restraints. The Atlantic salmon commercial fishery, long in decline, had essentially collapsed by 1860. One displaced fisherman, William Hume, crossed the continent with a homemade gillnet in 1852 and set it in the Sacramento River. Other East Coast fishermen followed Hume, seeking opportunities in western waters. These canny fishermen were able to obtain or thwart passage of regulations as their interest dictated. They openly defied laws they were unable to change. As ethnic differences broke down and market forces strengthened, this became the common pattern of San Francisco Bay/Delta salmon fishery "management."

The first West Coast salmon cannery opened on a scow moored near Sacramento in 1864. Within twenty years nineteen canneries were operating in the Delta area. In 1882 they processed a peak two hundred thousand cases of fish. (A case was forty-eight 1-pound cans.) On the North Coast, several salteries and canneries were operated by non-Indians. On the Klamath, Indian-caught fish were processed with Indian plant workers. Salted salmon from Humboldt Bay and Eel River ports, considered superior to the Central Valley product, found its way to markets in the Pacific Basin and New York.

Choice salmon for everyone. Early twentieth-century commercial seining operation on the Eel River near Rio Dell. Commercial salmon fishing on the Eel began in 1851. (Humboldt County Historical Society)

Hydraulic Mining Effects

The second factor affecting salmon and steelhead during this period was hydraulic gold mining, which began in the 1850s. By 1859, an estimated five thousand miles of mining flumes and canals were in use. Streams used by salmonids for spawning and nursery habitat were diverted indiscriminately for mining. Hydraulic cannons, up to ten-inch bore, leveled hillsides, so badly destroying stream habitat that salmonids could not survive. An estimated 1.5 billion cubic yards of debris (enough to pave a mile-wide super-freeway one foot thick from Seattle to San Diego) were sluiced into waterways and swept downstream, causing flooding and vast destruction of natural salmonid habitat as well as towns and farmland. The Yuba River bed near Marysville became a wasteland of rock piles, mud, and tangled masses of tree roots and branches two miles wide. Streambed elevations were raised—six or seven feet as far downstream as

Catches of salmon such as these were commonly taken from coastal
streams. The mottled appearance resulted from the fish lying on streambank stones after capture. (Peter Palmquist collection)

Sacramento. Extensive levees had to be constructed to protect
riparian lands.

Although hydraulic mining was banned by federal law in 1884,
the huge slug of mining debris severely impacted streams. Its effects can still be seen. Much salmonid habitat was permanently
destroyed. After the 1882 peak canning year, the river net fishery
experienced general decline, and the last commercial cannery in
the Central Valley closed in 1919. A smaller canning operation on
the Klamath was closed in 1933. Salmon stocks of the Central
Valley basin were stressed beyond their resilience—the natural
capacity to deal with environmental blows. The combination of
overfishing and wanton destruction of habitat had reached the
point at which the economic equation yielded less than zero to the
commercial fishery.

While the river fishery declined, commercial trollers, noting the
success of ocean sportfishers, began harvesting salmon offshore. By
1904, some one hundred seventy-five sail-powered fishing boats

Lateen-rigged felucca, a type of fishing boat used by early Italian, and to some extent Portuguese and Greek, fishermen on San Francisco Bay. (National Maritime Museum)

were operating out of Monterey Bay. During the early decades of the twentieth century, as gasoline engines replaced sail power and revolutionized the industry, offshore commercial salmon trolling overshadowed river net fishing. This move to ocean grounds was seen by state fishery experts as the major reason why depleted salmon stocks could not recover. Fishermen fished huge areas over longer periods and brought to market catches of immature fish. Ocean fishing raised a host of regulatory problems to be resolved.

Montereys. The boats pictured in the 1930s at Fishermen's Wharf, San Francisco, were powered by one-cylinder gasoline engines. The Monterey replaced the sail-powered felucca in the California commercial salmon fishery. (National Maritime Museum)

Early Restoration Efforts

By the 1860s the social costs of essentially unrestrained exploitation of salmon resources were becoming obvious, but for the next half-century an unexplained phenomenon masked reality: the abundance of surviving stocks fluctuated dramatically. In 1878 salmon from the Sacramento River (which often accounted for 70 to 80 percent of the state's production, as it still does) were so plentiful that a twenty-five-pounder cost the consumer only fifty cents. As noted above, the Sacramento River salmon pack in 1882 peaked at two hundred thousand cases. By 1890 that number was a minuscule twenty-five thousand—one-eighth as many. By 1907, numbers had built to a twenty-year high, then faltered, rose, and fell again. The pack ultimately reached a low of three thousand cases in 1919—less than 2 percent of the 1882 high.

The reasons for the good years were many. After hydraulic mining stopped, its immediate effects gradually began to lessen. The State Board of Fish Commissioners, riding a worldwide wave of interest in fish culture, was becoming a potent force in state fishery matters. Scientific study of fishery biology obtained strong governmental support. The Indian fishery produced less drain on the resource because Indians were driven off the rivers or their numbers were reduced by genocide or disease. Recurring sets of high runoff years provided abundant water for fish. It was even suggested that gravel from hydraulic mining improved fish spawning habitat in main river channels.

A number of possible reasons also explained the downside. Hydraulic mining was blamed initially, but by 1910 overfishing was considered the chief cause. These, however, were not the only reasons. Railroad construction in the upper Central Valley in the 1860s impaired salmon migrations on magnificent streams such as the McCloud; rock-blasting explosives also worked fine for killing fish to feed the nine thousand workers. After the demise of hydraulic gold mining, fish did not get back their water because miners' dams and flumes became farmland irrigation systems. Sea lions on San Francisco's Seal Rocks were blamed. Diverse ethnic groups pointed the finger at each other. Salmon protection laws were disregarded by stalwart citizens, including law enforcement officers, who resented state interference in local affairs. Polluted runoff from rice fields, fish diseases, sawdust from streamside mills—all were suspect. Debris dams built to contain the downstream creep of mining wastes often became the first barriers to upstream migrations of fish on larger rivers. Because environmental science was in its infancy, other unsuspected reasons may also have been at work, among them oceanic conditions, transplants of fish to nonnatal streams, or the effects of widespread introduction of exotic species like striped bass, shad, and carp into Central Valley rivers.

In 1871 the American Fish Culturists Association, imbued with Darwinian zeal, persuaded Congress to establish a federal Fish Commission whose principal duty was to restore the nation's depleted fisheries. First among its California restoration activities was the establishment of a salmon hatchery on the McCloud River in 1872, from which eggs were to be transferred east to restore the Atlantic Coast salmon fishery. This measure soon led the California Fish Commission to initiate an ambitious state hatchery program. Also vigorously pursued was the passage of complex, ephemeral laws dictating how, when, and where fishermen could fish.

Enforcement problems persisted, with predictable effects: law-abiding fishermen suffered while lawbreakers caught more fish. In 1893 this problem was eased somewhat by the introduction of a commission-manned patrol boat, the *Hustler,* which prowled the Sacramento River day and night. Changes in court jurisdictions also helped. Despite such efforts, California's salmon resources early in the twentieth century continued to decline. The total commercial catch in 1880 was eleven million pounds; by 1922 it had dropped to seven million pounds; after great fluctuations it reached a low of less than three million pounds in 1939. Salmon stocks were depleted. Most fishermen had moved to better fishing grounds, but enough fish were left to support a marginal fishery—thus ensuring that stocks could not recover on their own. That was the classic "fisherman's problem" described by McEvoy.

Salmon and Water Interests

Agricultural water diversions have been devastating to fish life. Although farmers and fishermen were allies against miners, as that conflict ended the relationship changed. A law enacted in 1870 that required fishways around dams was never effectively obeyed, even though subject to periodic strengthening amendments. Presently, as Section 5937 of the California Fish and Game Code, it specifies in essence that water diversions must not interfere with fish migration. Nevertheless, many still do so. Section 5937 would seem unarguable; however, water diversion remains today as the most crucial issue facing anadromous salmonid fishery restoration efforts. Central Valley irrigation issues are the core of the problem.

The Central Valley Project (CVP)

While commercial fisheries faced serious problems at the turn of the century, plans to develop California's water resources, chiefly for flood control and farm irrigation, received close public attention. Five devastating floods during the years 1902 to 1909 led the legislature to seek federal help. The U.S. Army Corps of Engineers, active on California's water scene since 1868, conducted extensive debris removal projects in Central Valley rivers and, at state request, developed flood control projects in the Central Valley. The Bureau of Reclamation, a federal agency that has made its mark on the West by building dams and distributing water, began studying water development potentials in California in 1905, focusing on reclamation of agricultural land.

The final result would be extensive development of irrigated agriculture in California. Both the corps and the bureau sought lead roles in that endeavor, but the bureau ultimately was placed in charge. In 1919 a proposal for comprehensive statewide irrigation development, the Marshall Plan, won lawmakers' approval. By 1931 this plan had evolved into an undertaking later called the Central Valley Project. State bonds could not be sold during the Depression, so California again sought federal help, and, in 1935, emergency relief funds became available for the project.

The Bureau of Reclamation began building Contra Costa Canal, the first unit in the vast project, in 1937. Friant and Shasta dams, southern and northern mainstays of the plan, obstructed fish passage in the early 1940s. Friant extirpated the already impaired San Joaquin River as a viable salmon stream. After construction contracts were let for Shasta, the Bureau of Reclamation commissioned a blue ribbon panel of fishery experts to study possible problems and propose solutions for salmon when they butted against concrete barriers. The resulting document, the Shasta Salvage Plan, provided for a broad range of structures and procedures to save the several stocks of Sacramento River chinook salmon. Heavy reliance was placed upon natural river propagation. Of many elements in the plan, only the fish trap at Keswick Dam, below Shasta, and Coleman National Fish Hatchery were successfully implemented and remain in operation today. Other elements were not started, were started but never completed, or were completed but did not work.

Irrigation, power production, and flood control were the major purposes of the Central Valley Project. Maintenance and improvement of fishery resources were among several subordinate purposes. While the postwar stage of the project was taking shape in the mid-1940s the Bureau of Reclamation's comprehensive planning documents, published in *The Central Valley Basin*, touted the "highly successful" Shasta salmon salvage program as a model for the future. Potential effects on fisheries were considered in discussions of several project elements. A letter of transmittal from Secre-

Best intentions weren't enough. High water in December 1941 washed out the fish trapping facility at Balls Ferry on the Sacramento River. Rebuilt in 1943, it was abandoned as unworkable within three years. This structure was part of the "Shasta Salvage Plan" to relocate salmon displaced by the construction of Shasta Dam. (Coleman National Fish Hatchery)

tary of the Interior Julius Krug to President Harry Truman gave assurance that "every effort will be taken to preserve fish and wild-life resources."

Many dams were seen as potentially favorable to fish life. The federal Fish and Wildlife Coordination Act (1934) to integrate water development with natural resource preservation ostensibly was in effect. Rosy predictions were made for future salmon runs on major streams affected by the project. If adequate fish protection measures were taken in planning and operation of project facilities, salmon runs then estimated at 250,000 could be increased to 640,000—a whopping 156 percent gain! The team of federal and state fishery experts making that prediction emphasized, however, that while potentially devastating effects of dams on anadromous fish runs were well known, the efficacy of proposed salvage and enhancement plans had yet to be proved. At least four years of planning time were needed before construction.

This requirement conflicted with the bureau's "first-things-first" philosophy. While engineers enjoyed the luxury of solving their problems on paper beforehand, fishery experts were required to apply virtually untested concepts. Bureau policy was clear: water and power came first. Fishery protection measures would be included in project facilities if they could be accommodated with minimal effect on major objectives. The concerns expressed by fishery experts were duly reported but had little effect on construction or operation of the CVP.

To well-meaning BuRec functionaries trying to accommodate diverse constituencies, the process was routine: priorities had to be established; some special-interest groups had to accept less than they demanded. Wartime pressures to complete the Shasta power facilities, coupled with material and personnel shortages, diverted attention away from fisheries. Bureau policy thus made fisheries expendable. While national emergency restrictions could partly explain the bureau's earlier neglect of fishery values, that excuse was invalid. Had the bureau genuinely acknowledged fish and wildlife values, fish protection planning could have begun with preliminary engineering studies and been realistically paced throughout the planning process. That did not happen.

The federal Bureau of Reclamation is widely blamed for subsequent fishery declines traceable to the CVP, and with reason. The agency gave lip service to fishery conservation causes, but its actions belied its words. The Shasta Salvage Plan failed. The San Joaquin fishery died. Trinity River salmonid stocks have been decimated. In recent years bureau policy appears to be changing, but most likely cosmetically. The bureau's primary mandate to provide agricultural irrigation water precludes major improvement of fisheries.

The State's Role

Less well recognized is the state's role in Central Valley Project fishery issues. Fishery professionals of the state Department of Fish and Game and the U.S. Fish and Wildlife Service have worked together over the years to try to protect and improve fish resources. Their efforts have been seriously limited, however, by legal, political, economic, and social constraints.

Two events during the 1950s demonstrate the kinds of problems they encountered. The first was a federal court case, *Rank v. Krug*. In 1951, when farmers, duck hunters, and fishermen were suing the federal government to increase flows in the San Joaquin River below Friant Dam, California Attorney General Edmund G. ("Pat") Brown, at variance with a related finding of his immediate predecessor, declared that "the United States is not required by State law to allow sufficient water to pass Friant Dam to preserve fish life below the dam." The federal government was exempt from "state interference."

The second, closely related, event was also a blow to fishery interests. It involved the June 2, 1959, decision (D 935) of the California Water Rights Board, which declared that salmon protection on the San Joaquin River was "not in the public interest." This 1959 decision merits closer examination because it further reveals the collaborative nature of state and federal political mechanisms involved in both cases. In the 1930s and early 1940s, when Friant Dam was planned and constructed, the Bureau of Reclamation had not acquired water rights related to that project. During the late 1950s, when the bureau initiated the process of establishing those rights, state Fish and Game staff decided to reopen the legal question involving water for salmon in the San Joaquin River. The department, through Jack Fraser, its chief of the Water Projects Branch, filed protests with the Water Rights Board to justify its claim. That move stirred the water development community to action at every level.

First, the attorney general declined to represent the Department of Fish and Game, claiming that his office foresaw conflict of interest problems if it were asked to represent both the Department of Fish and Game and later, if there were appeals, the Water Rights Board. After much wrangling and foot-dragging, the attorney general authorized the department to employ outside counsel to represent it at the hearings but then balked at providing adequate funding for the purpose. Fraser searched intensively and finally settled on Wilmer W. Morse, a former deputy in the attorney general's office at Sacramento and recognized expert in judicial review of administrative procedures. Realizing that an adverse decision by the Water Rights Board could be expected, Morse and Fraser concluded that their efforts should focus on laying the groundwork for a successful appeal.

During this period, the U.S. Fish and Wildlife Service sought vigorously to enter the case in support of state Fish and Game. The Department of Interior denied the service's request on the grounds that "it would not be in the best interest of the United States for an agency of this Department to support the position of the Department of Fish and Game."

In January 1959, Pat Brown became governor. He appointed veteran Bureau of Reclamation waterman William E. ("Bill") Warne as his director of the Department of Fish and Game. The June 1959 decision, as expected, was adverse. At that point, Morse and Fraser, with approval of Director Warne, devoted themselves to preparing an appeal (petition for judicial review), required to be filed within a thirty-day period. It would include an allegation that Brown and former Water Rights Board Chairman Henry Holsinger had acted to circumvent fish protection laws in 1951 when Brown as attorney general, with Holsinger's participation, had issued the "no state interference" opinion in favor of the bureau. Plans for the court action became known outside the department, and a delegation representing concerned Friant water users, the bureau, and the State Department of Water Resources insisted that Brown prohibit the Department of Fish and Game from filing the court action.

Just hours before the deadline, while mimeographed copies of the voluminous document were being stapled together, Governor Brown's secretary informed Morse that the governor had decided the case should not be filed. Morse insisted upon talking face-to-face with Governor Brown. The governor saw him at once, listened attentively, but declined to change his position. At the Department of Fish and Game, Jack Fraser, "furious as well as heartbroken," sought help from Director Warne, only to learn that "the door had been slammed shut, . . . William Warne was firm—the Governor's office had issued the edict, and that was it!"

Thirty years later, Jack Fraser recalls, "It was the low point in my career. I could not comprehend the crass circumvention of the public interest and destruction of public resources by purely political edict."

After serving for nine months as Director of Fish and Game, Warne was named state director of agriculture for a year. In January 1961, Governor Brown put him in charge of the Department of Water Resources.

In February 1989, UPI newsman Lloyd G. Carter reported an interview in which the octogenarian Pat Brown explained that he had always tried to balance the needs of salmon against other demands for water, adding, "We wanted to build the California Water Project and then work with the Bureau of Reclamation, and so I didn't want any confusion about it." (Funding for the State Water Project received voters' approval by a narrow margin in 1960.)

Despite Pat Brown's expressed intentions, the confusion engendered by those decisions of many years ago may not have been allayed permanently. The many complex issues of the bureau's Friant water contracts are being reexamined in a lawsuit initiated in 1988 by a consortium of conservation and fishermen's groups.

The Postwar Period
The Good News and the Bad

At the close of World War II, optimistic predictions for renewal of salmon runs in Central Valley rivers appeared to be coming true. Great volumes of cold water released from Shasta Reservoir were a boon to upmigrant adult fish, and ample flows assured their seaward-bound progeny safe downstream passage to the Pacific Ocean. Streams feeding the lower San Joaquin River were packed with salmon. In the upper Sacramento River, tourists at Redding Riffle watched the dark backs and thrashing tails of thousands of spawning chinook salmon. So crowded were successive runs of fish that new arrivals often uncovered redds of earlier spawners. Hungry steelhead and other fish consumed untold numbers of developing salmon eggs that were thus swept downstream.

Commercial fisheries prospered from this cornucopia. During the war, a cycle of favorable weather years promoted salmon survival, although fishing pressure increased. In 1939, commercial fishers had caught less than three million pounds of salmon; by 1945 and 1946 the number exceeded thirteen million. The commercial salmon fishing industry was coming back.

Sportfishermen also fared well. During the 1920s northern California's salmon and steelhead streams had earned worldwide acclaim, and the economic value of the sport fishery exceeded commercial fishing by two-to-one. In the mid-1930s ocean party-boat fishing began to gain public favor; between 1947 and 1955, that commercial-cum-sport industry skyrocketed, reporting 1955 landings of one hundred thirty thousand salmon—five hundred percent more than were reported just six years earlier.

It was indeed too good to be true. Large runs of salmon gathered at Redding Riffle in such abundance because impassable dams stood between them and their ancestral spawning grounds. Without choice, they spawned wherever they could. Of broader significance, irrigation contracts for Shasta water had not been let; thus, ideal stream conditions would exist for only a limited time. Future generations of those spawning fish would return to far less promising habitat and flows.

Water projects took on renewed life in the early postwar decades. In 1951, when snowmelt from Mount Shasta at last began watering farm fields in the San Joaquin valley, the legislature authorized studies for the State Water Project (SWP). Oroville Dam on the Feather River would be the system's key facility. The major purpose would be the transfer of northern California water south to Kern County and south coastal California for agricultural, industrial, and domestic use. Some two dozen other water districts up and down the state, thanks to their representatives in state government, also stood to benefit. Massive diversions would ensure that little water would "waste to the sea," an expression that became the shibboleth of SWP promoters. Construction began in the early 1960s.

An ironic note in the story of Sacramento River basin salmon: river gillnetting was banned in 1957, not for the welfare of salmon, but because of protests from striped bass fishermen. Gillnetters had to throw back incidental takes of stripers, and fishermen objected to seeing dead fish floating by, especially when their own luck was bad, so all river gillnetting was stopped.

Throughout the period the Bureau of Reclamation continued its Central Valley Project activities. In the mid-1960s the interbasin transfer of 90 percent of coastal Trinity River's water to the Sacramento basin was completed. The Corps of Engineers and the Bureau of Reclamation built dams on the American River. Hatcheries, with improved planning because of the strengthened federal Fish and Wildlife Coordination Act, were constructed at these diversions to mitigate lost fish habitat.

The Red Bluff Diversion Dam

In the mid-1960s the Bureau of Reclamation built the Red Bluff Diversion Dam on the Sacramento River. Working cooperatively with the state Department of Fish and Game and the U.S. Fish and Wildlife Service, the bureau incorporated an elaborate fish protection and enhancement facility into that irrigation project to mitigate lost spawning habitat. In lieu of a hatchery, the bureau constructed a state-of-the-art system of artificial spawning channels with accompanying fish screens, controlled water flow, and thousands of yards of ideal-sized spawning gravel.

At the dedication ceremonies, the late Vernon Smith, a spokesman for sportfishermen, happily told the festive group that this was a truly historic occasion. At last fishery and water interests, recognizing and honoring mutual concerns, had embarked on a positive, promising endeavor. The crowd clapped and cheered.

Citizen Action Begins

The first advisory committee. Commercial salmon catches after the first postwar years rollercoasted down by more than half, rose dramatically to twelve million pounds in 1955, then again plummeted to less than four million pounds in 1958. Erratic fluctuations marked the following decade's below-average catch figures. A host of factors—from economic forces to weather and ocean conditions, including possibly normal fluctuations in salmon populations— made the immediate causes difficult to pinpoint. The Department of Fish and Game concluded that the principal threat to salmon stocks was the declining survival rate of young fish and in 1968 announced a plan to reduce sportfishing bag limits, close some areas to fishing, and shorten the season. Further steps in 1969 would include curtailing the commercial fishery season and intensifying hatchery production and screening of major water diversions.

During that time, environmental activism was becoming an accepted American phenomenon. The California Environmental Quality Act (1970), the National Environmental Policy Act (1970), and later the California Wild and Scenic Rivers Act (1972) were among the dozens of legislative products of that movement. When the Department of Fish and Game announced its plan to further restrict harvest, alert fishermen-activists felt wronged. The department had emphasized that overharvesting was *not* the cause of the decline, so why should fishermen suffer the consequences? Commercial salmon fishermen were already being forced to sell their boats, a painful and unfair way to learn conservation imperatives.

Historically, government bureaucrats tend to see fishery management as a simple supply and harvest proposition, but salmon regulations always seem to emphasize restrictions on the harvest. Why not do something about the "supply" side—restoration of suitable natural spawning and rearing habitat? Under the leadership of William Grader, a Fort Bragg fish processor, commercial fishermen successfully petitioned the legislature to create a citizens' advisory committee to study causes of salmon and steelhead declines and recommend remedial action. The California Department of Fish and Game and the U.S. Fish and Wildlife Service agreed to provide consultant services.

The 1971 report of the Salmon and Steelhead Advisory Committee, titled *An Environmental Tragedy*, aimed at reversing the declines and insisted that the Department of Fish and Game place highest priority on restoration of natural habitat. (Today hatcheries contribute about half of the total salmon production.) The legislature was urged to assure adequate and equitable federal and state funding and to guarantee that future projects affecting the resources would include fish protection as a purpose. In a special section dealing with the Bureau of Reclamation's Trinity River Division, the committee noted that Trinity steelhead runs had declined 82 percent since 1961 and urged priority action to correct the faulty technology contributing to the failure.

The second report, *A Conservation Opportunity*, published in 1972, expressed satisfaction that the legislature and Fish and Game had already acted favorably on eight of nine committee recommendations aimed at restoring salmonid resources. The report then detailed opportunities for state and federal governments to take further restoration steps. Amendments to various federal acts dealing with water and power projects were addressed. The group also recommended that the federal government should make up fully for past neglect of fishery protections in Central Valley Project works.

The 1975 report of the Advisory Committee, titled *The Time Is*

Now, emphasized the need for local initiative and suggested ways in which state and federal governments could contribute to this effort. Reporting disease problems and other failures of hatcheries (the Nimbus Hatchery had just lost seven million fingerlings to disease) and expressing concern that the Red Bluff Diversion Dam spawning channels were not producing hoped-for results, the committee reemphasized an earlier suggestion: development of off-stream rearing ponds.

Hatchery operations often produced surpluses of young salmon and steelhead that had to be released directly into streams, where few survived. The citizens' group suggested that these fish be made available to local communities for nurture in offstream rearing ponds, then released into streams to migrate freely to the ocean. The concept had been tried on California's North Coast, where it generated keen interest among conservation organizations, industry, and the general public. The department liked the idea and volunteered to provide technical expertise. The legislature provided start-up funding and later established more substantial fishery restoration accounts. Offstream rearing projects have since expanded on coastal watersheds with promising, but debatable, results. Fishery experts raise serious questions about the genetic implications of the practice.

The Upper Sacramento Committee.   Despite follow-up action related to the citizen committee's recommendations, salmon runs in the Sacramento River continued to decline. In 1982, Charles Fullerton, director of the Department of Fish and Game, appointed a citizens' advisory committee to explore salmon and steelhead problems on the upper Sacramento River, above the mouth of the Feather River. Salmon runs there had declined since the 1950s from four hundred thousand to fewer than one hundred thousand, and the percentage loss of steelhead was even greater. The Upper Sacramento River Advisory Committee has been characterized as a rifle aimed at specific problems, compared with the shotgun approach of the legislature's committee, which studied statewide conditions. Both committees benefited from expert consultant services provided by state and federal fishery agencies.

The first target of the Upper Sacramento Committee was Red Bluff Diversion Dam. In a well-documented report, the committee labeled that facility as "perhaps the single most important cause" of the declines. The dam seriously obstructed upstream spawning migrations, caused the destruction of millions of seaward-bound juvenile fish, and, despite determined efforts of state and federal fishery planners, the mitigation facilities simply did not work. The committee recommended several steps to correct the migration problems, but it saw little hope for the artificial spawning channels. Abandon them, it concluded, and rework part as rearing ponds for juvenile fish.

Subsequent studies of the Upper Sacramento Committee dealt with problems of Coleman National Fish Hatchery and an Army Corps of Engineers bank stabilization project between Chico and Red Bluff. Their fourth report, in 1986, focused on the Glenn-Colusa Irrigation District's faulty fish screen operation, a long-standing problem that causes downmigrating fish losses estimated to equal the production of Coleman Hatchery. In essence, the committee recommended that federal and state agencies, employing existing legal mechanisms such as Fish and Game Code Section 5937 and the public trust doctrine, should force the water district to correct faulty conditions. If that did not work, the GCID gravel dam should be removed and pumping should be reduced or eliminated entirely. The committee's tough position is supported by a host of conservation-minded fishery groups, including the United Anglers of California, a politically powerful group representing thousands of various sportfishermen. The matter is currently being debated at state and federal levels.

The imminent possibility that the winter run of chinook salmon in the Sacramento River will become extinct commanded the attention of the Upper Sacramento River Advisory Committee in 1988. This unique race, its roots intertwined with fish that railroad crews blasted in the McCloud River in the 1860s, had declined in numbers from one hundred and seventeen thousand spawners in 1969 to fewer than twelve hundred fish by 1980 and remains at a precariously low level. Actions by the American Fisheries Society and a local conservation organization to have this run declared a threatened species ultimately led to development of a joint state/federal restoration plan. The Advisory Committee evaluated this plan in 1988. While recognizing its potential strengths, the committee decried the lack of progress in putting it into effect and determined that the planned program was legally unenforceable and administratively weak. The committee's recommendation: the winter-run chinook salmon of the Sacramento River should be listed as endangered or threatened under both federal and state endangered species acts.

Current Indian Fishery Issues

Citizen action related to salmon during the later postwar period extended beyond traditional sport and commercial fishermen's concerns. North Coast Indians were keenly aware of potential tribal benefits inherent in the environmental movement. Encouraged by their respected elders, they reminded non-Indians that the economic well-being of tribes in the Klamath/Trinity watershed had been irreparably damaged by the 1933 closure of the Klamath River commercial canning operation. They determined that such a debacle must never recur.

Although the Hupa and Yurok tribes had for some years experienced unresolved internal strife over timber sales, fishery issues unified them with the upriver Karoks toward one goal: economic development of the Klamath basin, with strong Indian voice in management. Since the Central Valley Project's Trinity River Division had hurt both river and ocean salmon fisheries, non-Indians shared Indians' concerns, but they did not welcome possible Indian control of the fishery.

During the 1970s Indian determination to realize their goal led to a series of state and federal court decisions that in effect established Indian rights to fish in traditional ways and compete with non-Indians in the commercial salmon fishery. With the passage of the federal Fisheries Conservation and Management Act in 1976, strife between Indian and non-Indian commercial fishers reemerged.

Results

Governmental positions relative to fish and wildlife issues from the early 1970s to the present have switched noticeably for a variety of reasons. First was a growing popular awareness that effective government requires citizen involvement. The 1950 *Rank v. Krug* case was often cited as an egregious example of governmental giveaway of natural resources. Its sequel, the 1959 stance of the state Water Rights Board that San Joaquin salmon conservation was not in the public interest, seemed unbelievably callous. Even the Bureau of Reclamation, structural specialist, was discussing "nonstructural" solutions to fish and wildlife resource problems.

Recent decades are notable for a plethora of governmental actions potentially beneficial to salmonid resources. Foremost was passage of the federal Fisheries Conservation and Management Act of 1976 (Magnuson Act), which extended federal control of the nation's fisheries from three miles offshore to two hundred miles, inaugurating an innovative system of federal/state fishery management. A new entity, the Pacific Fisheries Management Council (PFMC) with representatives from California, Idaho, Oregon, and Washington, was created to mediate fishery matters among user groups. Fish from the four states commonly range up and down the Pacific Coast and are targets of various commercial and sport fishermen. State governments had long dealt with resulting interstate problems and still maintain authority to three miles offshore. The Magnuson Act, through federal fishery agencies, gave the PFMC authority to resolve interstate and federal ocean fishery problems in a representative forum. PFMC's biological staffs also introduced computerized "preseason abundance forecasts" that missed the mark by as much as 300 percent and led to regulatory snarls involving allocation agreements between Indian and non-Indian commercial fishers harvesting Klamath River stocks.

Although its administration is still far from perfect, supporters of the Magnuson Act point to two of its benefits: better control of foreign fishery competition and a comprehensive resource-oriented approach to ocean fishery management that will, in the opinion of many professionals, ultimately benefit fish and fishers alike. In the meantime, the offshore troll fishery has been seriously affected.

Inclusion of several rivers, principally along the North Coast, in the California Wild and Scenic Rivers System (and later their further protection under federal law) was a widely accepted victory for conservation-minded Californians. Such actions affirm growing public recognition that maintenance of fishery and other wildlife values qualifies as the highest use of these aquatic resources. Among other encouraging governmental acts of the period were passage of the Forest Practice Act to lessen environmental damage from logging

(1973); announcement of the state Resources Agency's goal to double salmon and steelhead stocks (1978); establishment of the Renewable Resources Investment Fund (1979); the California Environmental License Plate Fund (1979); inauguration of the Salmon Stamp Program, a highly successful commercial fishermen-funded salmon and steelhead restoration program (1982); and coordination of operating procedures for federal and state water projects (1985). Also introduced were a program limiting new entries in the commercial salmon troll fishery, completion of a fishery agreement between the state and Indian tribes, strengthened provisions for fish and wildlife protection in water project planning, and a number of additional funding bills.

At the federal level, Congress in 1984 authorized a $57 million ten-year Klamath/Trinity fishery restoration project, and a total of $2.5 million (about one-tenth of the needed amount) was provided for renovation of Coleman National Fish Hatchery. Improvement of fish passage facilities at Red Bluff Diversion Dam were also funded; although initial results were encouraging, detailed evaluation in October 1988 revealed that major additional structural and operational improvements were needed to ensure lasting benefits.

Less popular among fishermen was the 1972 passage of the Marine Mammals Protection Act. Seals and sea lions intercept upmigrant adult fish at river mouths and tear hooked salmon off fishermen's lines in the ocean. Fishermen see this increasingly as a threat to their fisheries. Frontier-style controls are no longer tolerated, and populations of these marine mammals are expanding rapidly. Conservationists point out, however, that this relatively minor "threat" must not be allowed to divert attention away from bigger problems, such as loss of habitat that may cause total extinction of some species.

In 1988 a citizens' initiative, The Wildlife, Coastal, and Park Land Conservation Act (Proposition 70), received voters' approval. This act provides $16 million to restore, enhance, and perpetuate salmon, steelhead, and wild trout resources throughout California.

Perhaps the most significant governmental act of the period, at least potentially, was the Racanelli Decision in June 1986. The following description of this monumentally important court decision is offered by William Davoren, executive director of the Bay Institute of San Francisco:

Named after P. J. Racanelli, the presiding justice of the First Appellate District, State Court of Appeal, the decision was accepted by the State Supreme Court in September 1986. The case grew out of appeals of the State Water Resources Control Board's 1978 water rights order (D. 1485) and water quality plan for the Delta and Suisun Marsh.

Racanelli established that the State Board must consider all water users of the Sacramento/San Joaquin system, not just the big federal and state projects, and called on the Board to apply public trust doctrine principles as established by the Supreme Court in the Audubon–Mono Lake precedent.

The decision sharply criticizes the Board for giving too much consideration to water rights and failure to give enough consideration to its responsibilities for protecting fisheries and water quality. The timing of the decision seriously affected the Bay-Delta Hearing, now under way, that is scheduled to end in July 1990 with a new order and water quality plan replacing the interim standards of 1978.

Too Much Help?

Legislation and other public support during recent years introduced an ironic element: fish and wildlife offices were unprepared to deal with the new opportunities. It has been a period of confusion. The Department of Fish and Game developed no comprehensive salmon and steelhead management plan because it lacked sufficient data about the resources and was unable to hire needed personnel because of a hiring freeze in state government. No coherent plan for evaluating outcomes of restoration efforts existed. Overtaxed enforcement personnel were unable to monitor compliance with new laws. So many public bodies had an interest in salmon that interagency coordination of planning and restoration efforts had to be crafted on an ad hoc basis.

In 1983, California's governor, George Deukmejian, announced his intention to appoint as new Fish and Game director a woman considered by environmental groups and some legislators to be obviously unqualified for the position. Although this action was soon rescinded, it signaled the governor's lack of concern about natural resource conservation. Water developers sensed renewed freedom, and federal water brokers, advertising surpluses, sought additional Central Valley Project customers.

Also that year, a natural disaster compounded fishermen's woes. A periodic air and oceanic condition called El Niño developed. The upwelling of Pacific Ocean waters that supplies rich nutrients to make California fishing grounds so productive suddenly ceased. The ocean became eerily clear, salmon all but disappeared, and the few that were caught were as thin as snakes. The commercial catch dropped to the lowest ever—less than 2.5 million pounds, down 70 percent from 1982. Nineteen eighty-three was a bad year for salmon and the salmon fisher community.

Recent Citizens' Actions

Fishery leaders, in a forum (now the annual "Fisheries Forum") established by the legislature's Joint Fishery and Aquaculture Committee, called for reestablishment of the Advisory Committee on Salmon and Steelhead Trout. The legislature concurred and charged the committee with responsibility for developing a comprehensive statewide plan for management of these resources.

Representatives of commercial and sport fishermen, fishery biologists, Native Americans, and the general public were appointed to the committee. With initial financial and staffing help from the Department of Fish and Game (and later full funding by the legislature), it divided the state into eleven basin study areas and recruited hundreds of local citizens and professionals to determine the status of salmonid resources and recommend restoration programs. Simultaneously it initiated in-depth studies of common problems such as lost and degraded habitat, economic pressures, lack of public education, insufficient and polluted water, and a half dozen others.

The committee's work was reported in three publications: *The Tragedy Continues* (1986), *A New Partnership* (1987), and *Restoring the Balance* (1988). The earlier publications reported committee planning and progress toward meeting its charge. The 1988 report urged official state commitment to doubling of salmon and steelhead stocks and listed more than a hundred findings and recommendations pointing the way toward realization of that goal. The most urgent recommendations included strengthening the California Forest Practice Act and enforcement of streamflow requirements for fish and wildlife. Also emphasized was the need for legislative action to halt further federal water marketing efforts until the current Bay/Delta water quality studies by the State Water Resources Control Board are completed.

The Advisory Committee also recognized the urgent need to build public awareness of its concerns. Toward that end, it financed initial development of a public school education program and a professional-quality videotape, *On the Edge,* focusing on salmon and steelhead, and encouraged preparation of this volume. It also developed a plan to utilize the legislature's resources to make the committee's voluminous findings and reports publicly available at low cost.

The 1988 report led to several legislative initiatives, the most significant being SB 2261, the California Salmon, Steelhead Trout, and Anadromous Fisheries Program Act. This legislation required the Department of Fish and Game to "prepare and maintain" a comprehensive salmon and steelhead conservation and restoration program, coordinating its activities with the Advisory Committee. Doubling of California's salmon and steelhead stocks by the year 2000 was declared to be state policy. Funding included $250,000 for Department of Fish and Game start-up costs. During the fall of 1988, the Advisory Committee began defining its monitoring responsibilities relative to the plan; in 1989, it began working with the Department of Fish and Game toward its implementation.

In other ways, as well, the "winds of change" that Stuart Udall, a former head of the Department of Interior, heard blowing during an earlier administration were clearly felt during 1988 and 1989. In 1988, the California commercial salmon fishery landed a record 14,671,400 pounds of salmon. The ocean sports catch exceeded 200,000 fish, slightly below 1987 figures. Other "good news" was the determination by federal and state authorities that the Sacramento River winter chinook run qualified as an endangered species, thus deserving of extensive, not yet fully determined, protective measures. The State Water Resources Control Board issued a draft plan for control of Delta water quality that included provisions favorable to fish life and suggested the need for a statewide "water ethic" that could necessitate a definition of "reasonable use" favorable to fish and wildlife resources. The Bureau of Reclamation acknowledged water temperature problems at Shasta Dam and announced development of a plan to provide plastic curtains to help cool water releases. The problem-plagued Red Bluff Diversion Dam spawning facilities were closed.

These developments were tempered, however, by a number of sobering facts: in 1989, a precariously low number of Sacramento River winter chinooks—five hundred and seven—returned for spawning at Coleman National Fish Hatchery. The eggs of only one pair were successfully hatched; some fish, however, spawned naturally in the river. The cost of installing controversial water temperature control curtains at Shasta Dam, initially estimated at $5 million, which Congress appropriated in 1988, was recalculated at about three times that amount. A more conventional—and more expensive ($50 million)—temperature control system recommended by engineers is now being considered. Questions of who should pay for it—the Bureau of Reclamation or fish and wildlife agencies—are being debated.

Klamath Management Zone Problems

Problems within the Klamath Management Zone (KMZ) continued, and commercial troll fishermen were restricted in 1988 to a total of 40,450 fish. That harvest quota was reached within a few days of season opening, whereupon the KMZ commercial troll season was closed and the sport fishery bag limit was reduced to one fish. A distressing "Catch 22" situation developed in that fishery, as it became apparent that successful efforts to improve conditions for Klamath salmon survival led unintentionally to the lowering of harvest quotas. The reason: the increased numbers of Klamath fish in the mixed-stock ocean fishery increased the "Klamath contribution rate," a key element in PFMC's determination of harvest quotas in the KMZ fishery. Thus, the greater the percentage of the protected Klamath fish in the zone, the lower the total number of fish available for harvest in the entire zone. Because Klamath salmon mix with salmon in ocean zones north and south of the KMZ, restrictions were also placed on those fisheries; the impact was felt by Fort Bragg fishermen and those farther south, who harvest principally salmon orginating from streams entering San Francisco Bay.

The 1989 KMZ troll fishery—with a 30,000 fish quota—was open for a longer period, but there was a twenty-fish-per-day limit to discourage larger "trip boats" from competing inequitably with smaller "day" boats. Fishing south of the zone initially was so poor, however, that the trip boats moved north and fished the Klamath zone anyway. The 1989 commercial troll catch dropped to five million pounds of salmon.

Because of complexities such as these, commercial trollers balked at spending "Salmon Stamp" money to support Klamath River artificial propagation improvements. Moreover, the trollers claimed that contribution rate figures revised upward by the PFMC violated preseason agreements. The state Advisory Committee, meeting in June 1989, scratched their heads over what should be done about a situation in which, as one member termed it, "more is less."

Water Issues

Nineteen eighty-eight was a second consecutive drought year, and until March 1989 a third such year appeared to be in the making. Because of this and other coinciding factors, control of water resources dominated California fishery and other environmental news during the period.

SWRCB Bay/Delta Salinity Plan Development

The Bay/Delta water quality issue before the SWRCB reached crisis level in January 1989, when the board's Phase II draft plan released just two months earlier elicited strong opposition from water development interests, who vowed to "fight this every way we can."

The draft plan was a disappointment to fishery interests as well because it lacked fish flow assurances for dry years and promised little hope for recovery of San Joaquin River salmon runs. It did include provisions for increased spring flows and reduced Delta pumping, however, and was grudgingly accepted. Central Valley and southern California water interests objected to the draft plan because of the fishery considerations and other reasons, and they succeeded in disrupting the board's schedule for development of a final plan.

In January 1989, the SWRCB announced a six-to-twelve-month

delay in the hearing process to provide time for further studies. The board also removed the increased spring flows from consideration in Phase II of the hearings and sought additional input from interested parties. In April 1989, the SWRCB announced that a revised draft plan would be released for public review in late 1989 and offered assurances that the revision would consider the provisions of SB 2261, the anadromous fish protection law sponsored by the Salmon and Steelhead Advisory Committee, to be an expression of state policy. Fishery activists anticipate that the question of increased flows and reduced pumping from the Delta, however, will remain contentious issues. Final elements of the plan are now scheduled to be completed by 1993. The original target date was mid-1990.

Renewal of Friant Water Contracts

Renewal of forty-year-old water contracts between the Bureau of Reclamation and San Joaquin Valley water districts also became a prominent issue. On December 21, 1988, the National Resources Defense Council (NRDC) and twelve (eventually thirteen) other environmental groups sued in federal court to force the bureau to review environmental impacts under the National Environmental Policies Act prior to the precedent-setting renewal of the expiring water contract with Fresno County's Orange Cove Irrigation District. Six days later, the bureau announced it would renew the Orange Cove contract, purportedly in keeping with 1956 reclamation law. The bureau's position, enunciated by the Department of Interior's solicitor, Ralph Tarr, was that since no major changes were involved, "mere renewal" did not require environmental review.

Following pleas from the NRDC-led group, and several members of Congress who asked President-elect George Bush to help resolve the matter, the Justice Department agreed to delay signing of the contracts for a month. At that point, the Environmental Protection Agency, unable during the prior year to convince the bureau that environmental reviews were required before contract renewals, requested the president's advisory Council on Environmental Quality (CEQ) to intercede. CEQ held hearings in Washington, D.C., and Fresno during April 1989 and on June 30 recommended that major environmental studies must precede contract renewals. It also explicitly expanded its review, recommending that the bureau conduct studies "for each of the water service units of the CVP prior to the renewal of the individual long-term contracts in that unit."

Federal District Judge Lawrence Karlton, hearing the NRDC lawsuit in Sacramento, initially enjoined the bureau from renewing the Orange Cove contract, but he later permitted temporary renewal pending outcome of the suit. In April, Secretary of Interior Manuel Lujan, declining to wait for CEQ recommendations, granted the Orange Cove contract and indicated he would renew another twenty-eight contracts before CEQ studies were finished. The increased price—from $3.50 to $14.84 per acre-foot—in the new contract would, he believed, "do more for conservation than all the courts." Several hundred more contracts will be expiring in coming years. CEQ's findings will influence the outcome of the suit. A court ruling favorable to the environmental groups could void all of the renewed contracts.

As suggested above, and confirmed by other recent developments, water economics—specifically water marketing—is coming to the fore as a key factor in California's ongoing struggle over control of water resources. The potential effects of this development on the state's salmon and steelhead stocks cannot yet be predicted.

Asian Driftnet Concerns

The Asian (Japanese, South Korean, Taiwanese) high seas driftnet fishery in the North Pacific also caused concern because of its possible effects on migrating California salmonid stocks. Careful study and monitoring of the operation has established that Northern United States, Canadian, and Soviet fish and wildlife populations have been seriously harmed by this fishery. Although California coho and chinook salmon stocks are felt to be unaffected because they generally range south of the driftnet operation, alarming evidence emerged in 1989 suggesting that California steelhead stocks are being hurt.

Richard J. Hallock, retired DFG marine biologist, prepared a study for the Upper Sacramento Salmon and Steelhead Advisory Committee which revealed that steelhead from at least five California streams—from Humboldt County's Van Duzen River to Carmel River on the Central Coast—migrate through the high seas driftnet area. A significant proportion of the state's steelhead population—as much as 10 percent annually—may be lost to that fishery.

Despite successful enforcement efforts of the National Marine Fisheries Service and the U.S. Coast Guard, resolution of the problem promises to be difficult because it impinges on international trade relationships. Federal law and international fishery agreements have only limited effect. In August 1989, the California Department of Fish and Game announced its participation in development of a coordinated coastwide plan to curtail the destructive effects of the high seas driftnet fishery.

A Future in Doubt

As California enters the final decade of the millennium, the status of its salmon and steelhead restoration efforts is at a critical point. Improvements are apparent: public funding is being provided; habitat restoration projects are progressing; public awareness is increasing. Long-range successes, however, will be crucially affected by legal developments such as the result of the Friant water contracts suit and the final outcome of a March 1990 ruling of the U.S. Court of Appeals in San Francisco that upheld the constitutionality of the Reclamation Reform Act of 1982. (The reform act restricted the use of subsidized water from federal reclamation projects to nine hundred and sixty acres per farm.) The initial effect of such legal developments is establishment of a level playing field. Until success is achieved in that realm, and ecological principles become broadly accepted, the future of California's salmon and steelhead will remain in doubt.

Chapter Two

The Klamath River Fishery

Early History

Ronnie Pierce

"The present is the living sum total of the past." This observation
applies directly to an understanding of the Indian fisheries of northern California's Klamath River. Two distinct pasts of the indigenous
river people, the precontact and the postcontact eras, have merged
to create the complex tangle of tribal values, economic realities,
political, legal, and environmental concerns that exists today. To
understand the current status of the Indian fishery, the controversy, and the confusion surrounding Indian fishing rights, one
must understand both of these pasts.

Precontact Era

The aboriginal territory of the Yurok people encompassed riparian
lands along the lower forty miles of the Klamath River, from its
confluence with the Trinity River, its major tributary, to the Pacific
Ocean. It also included coastal lands from a few miles north of the
river's mouth south to Trinidad. The Yurok people have lived and
fished on the Klamath River from time immemorial. The river was
their world. North, south, east, and west did not exist for them.
The only directions were upriver or downriver. The center of their
world was Qu'-nek, where the Klamath joins the Trinity.

As the Yurok world centered on the river, the peoples' lives
centered on its fish. Salmon especially, but also steelhead, lamprey,

Anthropologists termed it "primitive affluence." The chinook salmon held
by this Indian woman and boy were an important element in the coastal
Indians' "food bank." (Peter Palmquist collection)

and sturgeon, were the mainstay of Yurok subsistence. The awe-
some, cyclical nature of the salmon's yearly migrations over the
centuries influenced almost every aspect of Yurok life. Religion,
lore, law, and tribal technology all evolved from the Indians' relationship with their fishery resources.

Such dependence on salmon required conservation measures to
assure that the bounty would continue. The downriver Yuroks had
to let enough salmon escape to perpetuate future generations and
to meet subsistence needs of upriver tribes, the Hupas and Karoks.
These two needs, biological and social, for an adequate salmon
escapement were met with many rituals and laws that tempered
the salmon harvest.

To the aboriginal Yurok, fishing sites were (and to a great extent
still are) considered privately owned. The right to fish at a particular
site was transferable and governed by complex rules and laws.
Rights could be loaned for a portion of the harvest. Owners of the

Indian property. Elderly Indian couple at a "usual and accustomed" fishing site on the Klamath River. Such sites were held and controlled by Indian families. (Peter Palmquist collection)

best sites were generally the "aristocrat" families of the tribes. Fishing techniques varied with the topography of the site and the type of fish being sought. Several types of nets, woven with twine made from iris leaf fibers, were used for salmon. Private ownership of restricted net fishing sites, however, could not ensure adequate subsistence for all Yuroks, so communal fish dams were temporarily built at selected sites. These weirs, constructed of log frames and a latticework of slats or poles that completely crossed the stream, blocked upstream fish migration. Sections could be opened to allow fish passage. Trapped fish were removed by dipnets and other means.

Possibly the most advanced accomplishment of California Indian cultures was the fish dam on the Klamath at Kepel. Several hundred people were involved in the annual construction of this dam. Every aspect of its construction and use was highly ritualized: it consisted of exactly ten panels, was built in ten days, and was fished for only ten days. This community project ensured that subsistence needs of all river tribes would be met, and salmon runs perpetuated.

Weirs such as this on the Klamath River were commonly used to intercept migrating adult salmon. Yurok Indians observed complex rituals associated with the annual construction and operation of much larger community weirs on the lower Klamath. (Peter Palmquist collection)

Neither net nor weir fishing could begin until the "first salmon" ceremony took place at Welkwaw, at the mouth of the Klamath River. The tribal formulist, after complex ceremonial rites, would ritually "spear" the first salmon. No person could eat salmon until this ceremony was completed.

Aboriginal Klamath fishers faced basic fishery management problems like those of today: how to cope with natural fluctuations in resources, how to control harvest while maintaining a viable economy. The Indians needed fish to survive. Clearly, as historian Arthur McEvoy has noted, Indian communities, over the centuries, learned how to "balance their harvest of fish with their environments' capacity to yield them." Salmon runs continued. Settlers of the 1850s reported that the Klamath was "alive with the finny tribe." Such was the Yurok world when the first major immigration of non-Indians arrived at the Klamath River.

Postcontact Era

Except for meeting a few explorers and fur traders, Klamath/Trinity tribes had little contact with whites before 1850. The California Gold Rush changed that dramatically. Gold seekers started settlement of the region, and reports of hostile contacts between whites and Indians were increasingly heard. Settlers demanded that Congress resolve the "Indian problem." Department of Interior Agent Redick McKee negotiated peace treaties that would preserve a small portion of tribal home territories while ceding all other lands to the state. Indians were to be taught to become farmers to lessen their dependence upon game and fish for food. Their leaders were given a choice: Be peaceful and sign the treaties, or be killed or driven out of the country. At 4:00 P.M., October 9, 1851, Indian leaders signed the treaties.

The new Californians did not want reservations. The most vocal called for termination or removal of Indians, although there were no lands farther west where Indians could be transferred. An editorial in the *Los Angeles Star* of March 13, 1852, summed up the popular attitude: "To place upon our most fertile soil the most degraded race of Aborigines" on the continent and treat them "as powerful and independent nations, is planting the seeds of future disaster and ruin." A beleaguered Congress ultimately met in secret session in 1852 and rejected all eighteen California treaties. (The injunction of secrecy was not removed until a half-century later.) This action still clouds certain legal aspects of California Indian fishing rights.

Meanwhile, as white immigration increased, Indian lands remained unprotected and confrontations, often genocidal, increased. President Franklin Pierce finally, in 1855, established the lower Klamath River Reservation and set up a military post, Fort Terwer, "to lessen friction between the new Californians and the Indians." Early superintendents of the Klamath Reservation struggled to adapt the Indian people to an agricultural life-style on the rich lands of the estuary, but the Yurok fishers resented field work. They preferred to subsist on fish and native roots, berries, and seeds. They simply wanted to be left alone, unmolested by whites.

Indians' distress with forced agricultural labor was relieved dramatically during the winter of 1861–1862, when a flood destroyed the agency office and wiped out Indian dwellings, crops, and fields. Subsequently, the agent and staff were transferred north to Smith River. Although they were destitute, the Yuroks refused to relocate and were left to fend for themselves on the Klamath. This development, among others, led agency authorities to consider reorganizing all California Indian affairs. One result was the Four Reservations Act of 1864, which eventually established the Hoopa Valley Indian Reservation on the Trinity, upriver from the Yuroks. This putative abandonment of the Klamath Reservation was yet another glitch in the bureaucracy that in the future would prove a hindrance to the affirmation of Yurok fishing rights.

White immigration to Indian territory increased, and along with it the impact on salmon and other Indian fishery resources. Mining in headwater regions destroyed fish habitat, and miners drove Indians away from upper areas of the river. The conflict between Indian cultural fishing values and non-Indian industry had begun in earnest.

Predictably, confusion over reservation boundaries and locations led to further influx of squatters and increased pressure on government to remove the Indians and open the territory to homesteading. An army report in 1875 disclosed that the Yurok's main concern was their fish and predicted serious trouble if whites continued trespassing. Over subsequent years, squatters were ordered to leave, but the orders were largely ignored. A small military outpost was established at Requa to protect the Indians' fishing endeavors and attempt to maintain peace.

On April 1, 1876, the state of California legalized the sale of salmon caught in Del Norte, Humboldt, Shasta, and Mendocino counties. A non-Indian, Martin Van Buren Jones, soon established a commercial fishery at the mouth of the Klamath but was evicted by the army. He then moved his operation a mile up Hunter Creek, just beyond reservation boundaries. Since Indians benefited from this endeavor, the military did not object.

Thus the lower Klamath River Indians entered a new era. White men's dollars, instead of barter, paid for their harvest of fish and allowed them to earn a living as the white world required. Commercial salmon fishing was most suitable to Indians, as this was what they could do best. In 1886, John Baumhoff, owner of a saltery also on Hunter Creek, signed an agreement with twenty-six male

Yuroks who may be considered the founders of the first Indian fisherman's union. He agreed to provide nets and boats and to pay Indian workers ten cents for every salmon weighing over ten pounds. The participating Yurok fishermen agreed to fish for no other non-Indian operation.

Baumhoff may have anticipated incursion from another white entrepreneur. One Richard D. Hume, of the Atlantic Coast family who introduced salmon canning to California in 1864 and later moved to the Columbia, had applied unsuccessfully to the Department of Interior in 1883 to fish the Klamath estuary. In 1887, defying the governmental injunction, he brazenly entered the estuary with a small steamer outfitted to catch salmon, claiming that he was not on the land of the reservation but on navigable waters open to all. Brandishing a large-bore Henry Express rifle, he quickly won the first argument with the local military sergeant. Soon he brought in a large barge outfitted to salt fish and house his non-Indian crew.

Indian fishermen, who had labored to clear net-snagging debris from the river bottom, soon were in direct competition and conflict with Hume's fishermen. They requested governmental relief. Indian agents, ducking the navigable waters question, lodged criminal charges against Hume for unlicensed trading with Indians on the reservation. Large stocks of trade items and receipts of transactions with the Indians had been found on his barge, the first listed item being forty-eight pounds of Rising Star Tea.

In the case of *United States v. 48 Lbs. of Rising Star Tea*, the United States attorney did not appear, so neither the government nor the Yurok tribe was represented. Hume's attorneys won the case, the judge ruling that the Klamath River Reservation no longer had legal status due to its "abandonment" in 1862. In 1889 the U.S. Circuit Court affirmed the ruling. Although the Rising Star Tea case actually had little adverse effect on the Indian fishery at the time, it did add to future confusion over the reservation's status and Indian fishing rights.

During the period 1880 to 1891 no fewer than eleven different congressional bills were submitted to open the Klamath Reservation lands to homesteading. To resolve the matter, the federal government in 1891 extended the boundaries of the Hoopa Valley Reservation to include not only the first Klamath Indian Reservation on the lower river but also the land along the stretch of river connecting the two reservations, thus creating one reservation slightly larger than the two original reservations. That same year the California State Legislature decreed that the river was not "navigable water."

Congress seemed unaware of the formation of the enlarged Hoopa Reservation. In January 1892, just three months after the consolidation of the reservation, yet another homesteading bill, HR 38, was introduced. All lands of the original Klamath Reservation were to be subject to entry, settlement, and purchase by whites. Proceeds of the sale were to be directed toward removal, maintenance, and education of resident Indians. The House Committee on Indian Affairs concurred with the bill, arguing that allotments to the Indians prior to the public offer would be ill advised as the Indians "have no conception of land values or desire to cultivate the soil." (Little had changed on that subject over the preceding forty years.) The bill passed the House and Senate essentially as written, except that Indians residing on the old Klamath Reservation would be given allotments rather than be removed.

The ancient Indian village of Rek-woi (Requa by then), at the mouth of the Klamath, flourished under the dominance of white settlers. During the salmon runs, Indian people from upriver and neighboring territories came to work as commercial fishers and cannery employees. The few Yurok people who had survived wars, disease, and genocide moved into the twentieth century still situated on their aboriginal homeland and still closely tied to their life-sustaining resource, the salmon. But they were to face major assaults on that resource.

The question of navigability and the subsequent legal jurisdictional status and ownership of the bed of the Klamath River had been left unresolved in the Rising Star Tea case. In 1901, a murder was committed in reservation waters of the Klamath River, and the subject of court jurisdiction reopened the question of navigability. Upon admission to statehood, a state usually acquires sovereign ownership and control of all navigable waterways and the underlying beds. Thus the question arose in *Donnelly v. United States:* Should James Donnelly, the accused murderer, be tried in federal court for an act committed on a federal reservation or in a state court for a crime committed in "state waters"?

The fact that California had become a state prior to creation of the reservation in 1855 argued for state jurisdiction. But, on the other hand, California, on April 13, 1850, preceding statehood, had adopted English common law, under which all nonnavigable river-beds would be transferred to their proprietors, in this case the United States. Since the state had repealed the status of navigability of the river in 1891, the court held that the bed of the river was an integral part of the reservation and therefore under United States jurisdiction. Moreover, the judges found it absurd to consider that the reservation would be created to include the steep mountainous slopes on either side of the river but not the river itself, as it had been created for the Indians to "gain a subsistence by fishing."

Donnelly was convicted on other points, but his lawyers, in requesting a rehearing, focused on the riverbed issue. The court, maintaining that the river's legal status was peripheral to the case, recalled its position on that point and refused further oral argument.

Thus, with the question of title to the riverbed left unanswered and the status of the reservation left clouded by white settlement under the act of 1892, the state of California continued its presumed authority and regulation of all fishing activities on the lower Klamath.

Chapter Three

North Coast Salmon and Steelhead and Their Habitat (1)

Scott Downie

Edith Thomas guided her husband Joe's split-bamboo fly rod carefully through the thick brush shading the creek called East Branch. The last pool she would fish that July afternoon in 1910 lay before her. It was her favorite: deep and narrow at the top, it fanned out wide along the high blue rock marked by the kids' diving swing now adrift in the warm breeze. Below the blue rock the streambed gradually shallowed and formed a gravel shoal. In autumn she would be able to watch salmon dig nests and spawn in those gravels.

Cold water tumbled into the neck of this trout pool from the rapids above. Bubbles traced the deepest run, along the rocks and beneath a large fir snag suspended just above the water's surface. One end of the snag was root-bound to the steep bank, the other wedged behind the diving rock. Two of Edith's artificial flies decorated the log's thick bark. She'd lost them there three weeks earlier trying to cast far under the dead tree where she had seen an especially large trout rise to seize an insect struggling on the water, most likely one of the beetles or grubs infesting the log. The insects' decaying home had guarded the fish well that day; she had run out of tackle and time.

The water had felt too cold then for her to swim out and retrieve her flies. It was still cool, but the day was much warmer than on her previous excursion, and today she planned to get them after fishing through the pool. By now, last month's large trout might well have moved to another pool. Since these Eel River rainbows were really seaward-bound migrants, juvenile steelhead, Edith never doubted that another fish would have taken its place in the pool. Fish, even larger ones, were plentiful.

She and her fishing partner, Elmer Hurlbutt, the Thomases' best friend, had together plotted strategy to catch the dominant trout in this pool, as well as the many other pools they regularly fished during their monthly angling contests. Elmer liked to compare the fishing holes with his chicken coop: each with its established hierarchy. True, there seemed to be a mix of sizes, shapes, and a definite pecking order among fish, but Edith refused to identify these wild, robust creatures with his domestic barnyard fowl.

They were both correct; the steelhead trout in East Branch were highly specialized animals that demanded many unique conditions to accommodate their anadromous life cycle. They began life as eggs deposited in clean freshwater gravels. Within two months they hatched and then reared for as long as two or three years in their native creeks and rivers before "smolting" into the Pacific Ocean. There the juvenile salmonids quickly grew to adulthood and sexual maturity, which stimulated their spectacular return to the streams of their birth to repeat the cycle. After spawning, the steelhead often died. Thus was the cycle completed with this most gallant and romantic of fishes. The complex rigors of the salmonids' life cycles demanded environmental conditions found only in a few regions on earth.

These fish, however, along with their cousins, the salmon, had been thriving in the streams of northern California's Coast Range since the last Ice Age. Here the cool Pacific air pushed up over rising, steep mountains and produced heavy amounts of rain and snowfall. Over aeons, the young developing soils of the uplifting region were slowly distributed along newly forming stream courses. These forces eventually built flats, hills, and terraces that on this day in 1910 supported the coastal redwoods, among the largest, oldest lifeforms on the planet. This dense vegetation in turn helped to hold the area's fragile, steep slopes in a degree of stability, or at least in a condition of slowly metered erosion. The entire system, including the nearby Pacific Ocean, the winter storms and summer sun, the steep mountains and shaded streams, the oversized trees and over-leaping fishes, the myriad animals and insects, was therefore being

Fish fry in the making. A mixed bag of juvenile salmonids and sculpins (bullheads) taken from a Humboldt County stream. (Humboldt County Historical Society)

held in a state of tenuous equilibrium, a delicate balance, on that hot July afternoon during the first quarter of the twentieth century.

And so it was then into that vibrant pool that Edith Thomas and her Joe, followed by Elmer Hurlbutt and his Mary, their children, and then an entire wave of European immigrants crowded, to cast their neatly dressed artificial flies. The results were predictable.

By midcentury, after unprecedented decades of economic and technological development, the legacy of these early settlers and their progeny, in terms of exploited natural resources, was disastrously clear. The timber riches of the North Coast had spurred the highly mechanized harvest of vast tracts of Douglas fir and coastal redwood forests.

From the mid-forties to the mid-sixties, transient "gypo" loggers and sawmillers invaded the region with Gold Rush zeal. New and powerful bulldozers, trucks, and other heavy equipment developed during wartime helped fuel the boom. In most watersheds by

Promotional pictures are part of history. This woman models proper attire for a day of stream fishing in the 1920s. (Peter Palmquist collection)

the 1970s, more than 80 percent of the virgin forests had been cut, milled, and shipped south to eager lumber brokers, thence to consumers whose growing numbers, wealth, and demands seemed boundless.

But the fragile timberlands proved to be quite finite. In town after town, during the sixties and seventies, mills and businesses failed; populations, like the timber base itself, disappeared. The denuded watersheds, roaded and gullied beyond belief, had meanwhile suffered catastrophic flood events in 1955 and 1964. Unprecedented levels of eroded topsoil and mass wasting had left many areas no longer suitable for growing profitable conifer forests. These lands in turn became subject to sale and subdivision, which fostered a new land development boom in several North Coast areas that still continues. Once again roads and associated drainage problems were predictable elements of the new settlers' land use practices.

During all this frantic postwar activity the citizens of the region had reveled in the bounty of natural recreational opportunities as

Fishing holiday. These college-age sportfishermen pose in a redwood grove with their catch of Eel River salmon and steelhead. (Humboldt County Historical Society)

well. Many sources of outdoor enjoyment were near at hand: the seashore, the rivers and mountains, and abundant game. And, of course, fishing for the plentiful stocks of salmon and steelhead. It was upon these fish that the burden of years of poor use of resources first fell. Streams where the Indians, followed by the early settlers and loggers, had been able to spear or net salmon leaping from pool to pool soon had neither pools nor salmon, a result of flooded streambeds being choked with debris and sediments: a by-product of mindless timber harvest and land development.

Today many once pristine stream channels, like Edith Thomas's East Branch, lie buried beneath as much as fifty feet of gravel. Very few trees or shrubs shade these gravel flats during summer's heat. Consequently very little water is found in these streams after June; what little there is trickles through the stones and is soon too warm to support young salmon and steelhead except in a relatively few "cold spots." During winter's freshets the same streams often suffer disproportionately high flows induced by elevated runoff rates from extensive road drainage systems.

In addition to these direct soil disturbances, the reduced densities of large upper basin vegetation also encourage rapid runoff and prevent upper slopes from storing water for summer release. Add to this deadly scenario a liberal dose of stock grazing, hydraulic mining, wholesale irrigation diversions, intense fishing pressure, and high domestic water consumption, and it is easy to understand the 80 percent decline in salmon and steelhead populations that has resulted from the past two generations of modern consumptive mismanagement of these fragile, albeit renewable, resources.

Chapter Four

The Passing of the Salmon
Joel W. Hedgpeth

I will say from my personal experience that not only is every contrivance employed that human ingenuity can devise to destroy the salmon of our west coast rivers, but more surely destructive, more fatal than all is the slow but inexorable march of these destroying agencies of human progress, before which the salmon must surely disappear as did the buffalo of the plains and the Indian of California. The helpless salmon's life is gripped between these two forces—the murderous greed of the fisherman and the white man's advancing civilization—and what hope is there for the salmon in the end?

—Livingston Stone (address to the
American Fisheries Society, 1892)

Seventy years ago the chinook salmon of California was an important natural resource, as famous throughout the world as the gold, the redwood trees, and the city of San Francisco. With a prodigal disregard for the future, hundreds of thousands of pounds of salmon were taken from the rivers and canned for shipment to all parts of the world. The present-day salmon fisheries of the Columbia River and Alaska were nonexistent in the seventies of the last century—then the entire industry was restricted to San Francisco Bay and the lower Sacramento River. Any other than "California Salmon" was unheard of in those days. But that fishery did not last long. In less than twenty years it reached its peak and began to decline as quickly as it had risen. The canneries moved on to the Columbia, to Seattle and Alaska, and the words Alaska and Colum-

This essay originally appeared in *Scientific Monthly*, November 1944. Epilogue, July 1989.

bia River on the labels of the new cans became so familiar that most Californians forgot about their own salmon.

The canneries alone were not responsible for the decline of the salmon. It is possible that the intensive fishery between 1864 and 1882 had less effect on the salmon runs than the hydraulic mining that damaged hundreds of miles of rivers during those same years. Later, when the salmon were no longer considered an important resource in California, dams without fish ladders barred them from many spawning areas or held the water back from the riverbeds below, and unscreened irrigation ditches carried young salmon out in the fields by the millions to die. Yet all this was not enough to destroy the salmon completely. For years they have been coming back, trying to repopulate what is left of their rivers. But within the last few years man has devised new dams and water projects that will cut off much of the remaining spawning mileage from the salmon.

The year 1872 is a significant one in the history of the Sacramento salmon. In that year the newly established U.S. Fish Commission received an appropriation of $15,000 to be spent in the "propagation of food fishes." At a meeting held by Commissioner Spencer Fullerton Baird and attended by various New England fish commissioners and members of the American Fish Culturists Association, Livingston Stone, a retired minister who had recently taken up trout culture, suggested the importation of California salmon to replace the vanishing salmon of the New England streams. It seemed a good idea at the time, for it was not known then that the Pacific and Atlantic salmon are entirely different fish with radically different life cycles.

In the late summer of 1872 Livingston Stone and two young assistants found their way to the McCloud River nearing the end of their quest for spawning grounds of the chinook salmon. As they picked their way between the rocks and trees along the riverbank they looked into the water for signs of salmon. Many strange and rough characters roamed the hills of northern California those days—miners, hunters, surveyors, and less respectable individuals, but these three New England gentlemen in search of a site for a fish hatchery were a new sort. Stone, the retired Unitarian clergyman who sought outdoor work for the benefit of his health, was a stocky man, conspicuously shorter than his two companions. His

round head was framed by a pair of elegant brown dundrearies that
partially concealed his large ears. One of his companions was a
massive fellow whose solid chin was softened by a fringe of beard.
The other was slighter and beardless.

They did not have far to walk after leaving the ferry across Pit
River just below the fork where the McCloud comes in, and must
have been unprepared for the scene ahead of them. Two miles
upstream the McCloud turns toward the foot of a high limestone
crag and then makes another turn to the left along the front of the
crag. From the first turn the three men could see the gray rock
towering above the dense forest, the smooth water at the farther
bend, and the white churning of the water over the riffle at the
nearer bend. To their left was a sandy beach and a low hill on which
were clustered the brush huts of an Indian village. The Indians
were fishing, wading out on the riffle with long double-pronged
spears. On the other side of the river the forest grew almost to the
water's edge. The retired parson named the crag Mount Persephone, but on the maps it now bears the prosaic name of Horse
Mountain. The hatchery he built on the bank of the river within the
morning shadow of the crag he named Baird in honor of the first
commissioner of fisheries. Soon the new Shasta Dam will cover it
under nearly three hundred feet of water.

The three men lost no time in getting down to work after arriv-
ing on this scene. They had hoped to hire the Indians to help them,
but the Indians could speak no English. Working unaided in the
hottest part of the summer, the fish culturists built a house, a flume
for their water supply, and a series of hatchery troughs. The nearest
sawmill was at the railroad terminus of Red Bluff, fifty miles to the
south, and their lumber had to be hauled by wagon over the rough
mountain roads. In spite of these difficulties the job was finished in
two weeks, and on September 15 the first salmon were taken from
the river. The hatchery became a famous place in northern California as much for its cultured atmosphere—there were no oaths or
card playing, and its superintendent became the acknowledged
chess champion of the state—as for the queer things being done
with salmon eggs.

Fish culture has not advanced very far beyond the practice of
seventy years ago. Then a female was "stripped" of her eggs by

California's first fish hatchery. Livingston Stone and assistants at Baird Hatchery, near an Indian village on the McCloud River, c. 1875. A retired minister, Stone became interested in the emerging art of fish culture and tried unsuccessfully to transplant Central Valley chinook salmon to restore the East Coast salmon fishery. (Coleman National Fish Hatchery)

squeezing her somewhat after the fashion of milking a cow. Today she is hit on the head with a club and the eggs cut out of her body, which is less wasteful of the eggs. After the eggs have been removed, a ripe male is forced to ejaculate milt over them. The fertilized eggs are then placed in baskets in long troughs of running water. For the first week or ten days they are picked over for dead eggs. After ten days the eggs become "tender" and cannot be handled. Even jarring the tray at this stage may kill all the eggs. Then, about fifteen or twenty days after fertilization, the eye of the embryo appears as a small black spot on the egg. In this eyed stage the embryo is very tough, and the eggs can be packed in trays with moss, burlap, or ice and shipped to the far corners of the earth. They will hatch in six to nine weeks, depending on the temperature of the water in which they are placed. The only important change in this procedure has been the development of a drip incubator, in which the eggs are placed in shallow trays and water is percolated over them to cool them by evaporation. This method is not widely used, however.

The artificial propagation of salmon was so simple and obviously successful—after all, the fish were hatched—that it was assumed at the beginning that it was a notable improvement on the wasteful ways of nature. It is hard, in these sophisticated days, to realize the fascinated delight the early hatchery men took in their business of raising fish, how they watched over the eggs in the long troughs of cold water, picking them over with feathers like fussy hens, and the pride with which they watched the newly hatched alevins. They did not suspect that the natural hardihood and vitality of the eggs had as much to do with the success of their hatcheries as their human intervention in rescuing the eggs from the perils of the river.

That nature's methods of propagating salmon might not be as wasteful as they seem never occurred to them. Given half a chance, the salmon needs no assistance from man. It cannot, however, survive the total demolition of the rivers that is the inevitable result of power, mining, and irrigation developments.

During those early days of salmon culture in California no serious attempt was made to determine the efficiency of the natural spawning process. "In a state of nature, only two eggs in a thousand hatch" is the pontifical statement in one official bulletin. Livingston Stone dug into a nest and estimated that only "8 percent" of the eggs were fertilized. Although we still have much to learn about the life of the salmon, I suspect that natural spawning is at least as efficient as the hatchery method. Actual losses of eggs in hatcheries seem to range from 5 to 40 percent, with the emphasis in published records on the lesser figure. Nothing has been said of the possibility that hatchery-reared fry may not be as healthy as naturally spawned fish. Certainly the practice of dumping large numbers of young salmon into a river was more hazardous to the fish than their natural method of emerging from the gravel—aside from any acquired differences in their constitution. In recent hatchery practice this danger to the young fish has been overcome somewhat by permitting them to escape from rearing ponds in a "natural" manner.

Livingston Stone and his contemporaries had no misgivings of this sort about the efficacy of fish hatcheries. While it should not be forgotten that the life cycle of the chinook salmon was not completely understood at the time, their conclusion that the increasing fish catch was a demonstration of the value of the hatchery was a post hoc assumption of the first magnitude. They did not consider that greater fishing effort can increase the catch in a declining fish population.

In the beginning the hatchery on the McCloud River was simply an egg-gathering station. During the first season fifty thousand eggs were taken, of which thirty thousand survived to the eyed stage. These were packed in sphagnum moss and shipped east. In March of the following year, 1873, a few hundred fingerlings were released in the Susquehanna River. Thus began the unsuccessful attempt to transplant the Pacific salmon to the Atlantic, an effort that was not abandoned until ten or fifteen years ago.

Epilogue

The fishery investigations that inspired this essay occured in 1939–1940, fifty years ago. I was a junior member of the field crew assembled to work on the impact of the construction of Shasta Dam. During the long summer we were based at Baird on the McCloud River; in the winter some of us continued at Stanford University in a basement office in the museum, directly underneath Governor Stanford's locomotive No. 1. We often looked thoughtfully at the beams overhead, hoping that the next earthquake would not occur during working hours.

The report, by Harry A. Hanson, Osgood R. Smith, and Paul R. Needham, was published as Special Scientific Report No. 10 of the Bureau of Fisheries in 1940. One of its recommendations was that water be "drawn from the reservoir at a sufficient depth below the surface to provide cold water in the river below the dam for at least the first five or six years of operation of the dam." This was intended "to avoid the danger of warm water during the period when the fish are being transferred" to the hatchery and other streams. I do not remember that anyone was told that the dam was designed to make such low-level drawdowns difficult or impossible. In any event, concrete was already being poured before our fieldwork was begun, and it was too late for alterations. Now, in mid-1989, after fifty years and several more dams, there is to be an attempt to provide cold water by holding back water from Shasta Dam and

releasing it instead via the tunnel from the Trinity Dam as part of
the effort to save the now officially endangered winter run of five
hundred and seven fish (in 1989). This will cost more than $4.5
million in power sales alone. Nothing is said of the potential effect
on the sad remains of the Trinity River.

In 1939 we counted 16,108 fish in the winter run. That count was
probably only a small remnant of what it was in the days of Livingston Stone, but the current count—fifty years later—is 15,601
fewer fish than that 1939 number. While we were counting fish that
year at Redding, the WPA guide to California was published, with
its strange reference to Baird, "where salmon are propagated in the
California Caves on the banks of the river." In 1984 this book was
reprinted as a historical document without change, but there was a
disclaimer: "Some things have changed since 1939." Indeed. The
rivers that supported the salmon run are no more, and soon there
will be none of us to remember the McCloud River as it flowed by
the old dormitory of the abandoned fish hatchery at Baird that was
our field base, to hear the sound of its flowing and watch the sight
of the young salmon feeding on their way downstream, flickering at
dusk like a thousand points of light. The hatchery building had
been at the bend of the river, opposite the high gray cliff that
Livingston Stone called Mount Persephone, with its almost inaccessible caves (famous for remains of Pleistocene vertebrates) high
above the water.

At that time it was apparent that the water which would immerse
this scene would be used primarily for generating power; agribusiness (that unlovely word which took the culture from agriculture,
invented by a renegade entomologist turned pesticide salesman in
Fresno—where else?) would come second, and another purpose, to
prevent salinity from invading the Delta, would remain the last
when not forgotten. Already there were plans for the next great
project, Friant Dam on the San Joaquin River. I do not remember
the name of the agriculture professor at Hilgard Hall in Berkeley,
but I have never forgotten his astonishment and indignation when I
told him the people interested in fish were asking that water be left
in the river for salmon: "What are they going to do with that water,
waste it?" (He shouted the last words so loudly that he loosened his
dentures.)

That was the first time I had heard of the idea, nevertheless old in the West, that water left to flow freely to the sea is wasted. It is the doctrine of the seekers and users of water, pronounced in ringing words by Governor Earl Warren at the California Water Conference he had called on December 6, 1945: "In my opinion we should not relax until California has adopted and put into operation a statewide program that will put every drop of water to work." Nearly eight hundred people attended that conference, representing state and federal agencies concerned with irrigation, dam building, and reclamation, chambers of commerce, labor unions, farming associations, and all. Indeed, there were statements representing all but the official concern of the federal and state agencies responsible for fish and wildlife. Fish and Game sent no one to speak with authority on behalf of the salmon; their cause was left to the sportsmen (who could not speak with authority) while John Reber was an honored speaker, proclaiming his gospel that we should dam off the rivers and turn San Francisco Bay into a freshwater lake. The governor's hope that his conference would demonstrate "the greatest good for the greatest number" became instead a blueprint for the destruction of California.

The proceedings of that conference, an inch-thick, closely printed volume, have been forgotten (I doubt that anyone read all of it), yet it is startling to find this eloquent statement buried in even smaller type in the appendix afterthoughts:

Into this welter of selfishness, a still small voice is making itself heard. The voice of the salmon—one of God's creatures whose extinction is threatened by materialistic engineers who only see water as a sterile inanimate liquid. It is evident now that water is a medium in which life occurs, life which has an important bearing on human beings because all life on earth is interrelated. Control of water is control of many forms of life and of human beings dependent upon that life—and we are all dependent perhaps in ways we do not always realize or appreciate.

Planning of water control must be expanded to include all the life-supporting values of water. With this broadened outlook, instead of deciding on control simply because the physical facts of water occurrence and structural feasibility favor it, the biological consequences too will be taken into account. Biologists may show us that nature is pointing the way toward even greater benefits than have heretofore been realized. Certainly, it is presumptuous to belittle the centuries and aeons of experience on which nature has built her economy. Prudence would seem to dictate that her guidance be accepted and mankind's mastery of materials used to aid her if possible. With a readjusted and broadened outlook, which includes the wisdom of the biologists, it is not unreasonable to believe that even more successful harnessing of the forces of nature for the service of mankind will be achieved and wider benefits for all mankind provided.

The statement was written by Everett A. Pesonen, chairman, Fellowship for Social Justice, First Unitarian Church, Sacramento. That still small voice obviously has not been heard in the right places. At a hearing held in Sacramento on August 26, 1989, by New Jersey's Senator William Bradley, the water establishment, led by David Kennedy, director of California's Department of Water Resources and self-appointed archbishop of the "water people," began the litany for more structures, more water for the fields and an ever-growing Los Angeles. He was complacently reiterating the notion that water is there to be captured and distributed. The chorus of Water Establishment Stalwarts followed him, expressing hope for federal funding to help them service growth forever, that they could return again and again to the river. As for the selenium and other pollutants concentrated in the drainwater, their intentions were good, but obviously there was still hope for the San Luis Drain to dump it into the estuary of San Francisco Bay.

About their anticipation of getting enough water to accommodate the endless growth of Los Angeles, Senator Bradley remarked: "You have a philosophical problem."

We do, indeed.

Chapter Five

Remember the San Joaquin
George Warner

Historical numbers of salmon spawning in the San Joaquin River system probably equaled the size of runs utilizing the Sacramento River and its tributaries. The impact of water development, however, drastically reduced San Joaquin runs long before major losses occurred on the Sacramento. I have been unable to find any records of the size of the fall run on the main stem of the San Joaquin, but it is recognized that this run was the first to be eliminated by low flows and high water temperatures. On the other hand, the spring run remained in relatively good condition despite diversion dams and canals. Uncontrolled snowmelt runoff aided the upstream adult migration and simultaneously allowed juvenile salmon to migrate to the Delta before irrigation demands reached their peak.

Among early agricultural developments influencing the size of the spring run on the San Joaquin River was Mendota Dam, located in the Mendota-Firebaugh area. By the 1920s it was diverting water into three large irrigation canals. A few miles downstream another barrier, the sack dam (a seasonal dam made of burlap sacks filled with gravel and other material), supplied water to a fourth canal. Several other pumps along the river furnished water to riparian users. None of these diversions was screened.

Mendota Dam, a low concrete and flashboard structure, had a primitive pool and jump-type fish ladder extending downstream

This essay originally appeared in *The Caddis Flyer*, 1987, published by the Tehama Fly Fishers Club. Reprinted with permission. Epilogue, July 1989.

from the center of the dam. Since the entrance to the ladder was some distance below the barrier, fish had trouble locating it. Also, the pools were too short and the weirs too high. Many salmon did use it, however, and others succeeded in leaping over the dam by taking advantage of the hydraulic pattern of the flow over the crest. Minor fish passage problems occurred at the sack dam.

In the late 1930s the Division of Fish and Game recognized the magnitude of the problem facing the salmon runs in the Central Valley, and several biologists from the Marine Fisheries Branch were assigned to inland water project investigations. Their major objective was to determine the size and timing of the adult salmon runs in the various rivers as well as the downstream juvenile migration pattern. These studies were curtailed during wartime, but by 1946 attention was again given to problems such as screening the canals at Mendota. Here a fyke netting program conducted by Richard Hallock found significant losses of salmon migrants during the irrigation season.

In 1948, disaster struck. Friant Dam, a major unit of the Central Valley Project, had been completed and the Bureau of Reclamation assumed control of the river. Ignoring pleas from Fish and Game and local sportsmen groups, bureau officials diverted water desperately needed by salmon down the Friant-Kern Canal to produce surplus potatoes and cotton in the lower San Joaquin Valley. Only enough water was released in the river to supply downstream canals and some of the pumps. At that time there was no effective Wildlife Coordination Act, and the bureau argued that the enabling legislation for Friant Dam said nothing about protecting fish and wildlife.

Litigation over water rights followed and lasted through the 1950s. Attorney General Pat Brown and Governor Earl Warren, pushing for the development of the Central Valley Project, supported the bureau. The state's position was that water for agriculture was next in priority to domestic use, and salmon had no legal claim to San Joaquin water.

While negotiations continued, the Fish and Game crew, of which I was a member, tackled the almost impossible task of saving the 1948 spring run. Although time and funds were limited, it was apparent that the run would have to be trapped below the dry reaches of the river and transported somehow to the live stream above either the sack dam or Mendota Dam. It was apparent that large numbers of fish would be involved, since the 1946 spring run past Mendota consisted of fifty-six thousand salmon and the 1947 run numbered twenty-six thousand.

The 1948 spring run salvage plan for the San Joaquin included three elements: a trapping and loading facility, tank trucks for transportation, and a suitable release site that would bypass the dry section of the river. A hurried inspection of the San Joaquin from San Joaquin City to Los Banos revealed that salmon would be stranded a few miles above its confluence with the Merced River. A weir was erected across the river at Hills Ferry above the mouth of the Merced so there would be enough streamflow to allow salmon to reach that point. The site had a number of advantages such as a fairly level streambed and vehicle access.

The weir, two hundred and fifty feet long, was constructed of stretched mesh netting suspended from a steel cable and anchored to the bottom with logging chains and sandbags. Adjacent to the west bank, at the upper end of the weir, a rectangular trap was constructed of pilings, stringers, and vertical slats positioned to permit flow through the structure while still confining fish. A funnel of webbing at the downstream end allowed fish to enter but not to leave. A gate was installed at the upper end that could be raised to permit fish to be crowded into a collecting device. This device, which we termed a "bucket," was a large watertight metal box with a circular trapdoor in the bottom and a mesh cover. Twenty or thirty salmon could be handled at one time.

To complete our collecting facility, a mast and boom salvaged from a San Francisco shipyard were erected on shore. Also salvaged was a winch that could be operated with a power take-off on my Jeep. We could hoist fish and water in our bucket and swing it in any desired direction. We were able to borrow from Coleman Hatchery three tank trucks with aeration gear for transporting trapped fish. These trucks had been standing idle during World War II, and the tires were in sad shape. Otherwise they were fine for the job, since the bucket could be positioned over the tank hatches.

It was determined that the Outside Canal could be used as a release site. This canal originated behind Mendota Dam and passed within eighteen miles of Hills Ferry. Fish would encounter no obstacles if they were hauled and released there. Once they

The anatomy of extinction. This California Division of Fish and Game crew, adapting war-surplus equipment to fish rescue work, attempted to save the spring-run San Joaquin chinook salmon in 1948. The species became extinct when diversions from Friant Dam "dewatered" the San Joaquin River. (George Warner)

arrived at Mendota via the canal they could proceed upriver to the deep pools below Friant, where they could spend the summer.

All systems go! The salmon arrived and the plan worked perfectly except for occasional trouble with the trucks. The operation went something like this:

1. Position truck under boom and fill tank half full of water.
2. Start aerator and start Jeep.
3. Crowd salmon into bucket.
4. Hoist bucket and swing over tank hatch.
5. Release salmon and water into tank.
6. Add water to fill tank.
7. Haul to release site.
8. Count fish as they leave truck.

Besides hauling fish, the crew was constantly cleaning debris from the weir, answering questions from the public, and discouraging poachers who stood around plotting how to raid the trap or tamper with the weir at night. There was no such thing as an eight-hour day, five-day week, or overtime.

After 1,915 salmon had been hauled successfully, disaster struck again. A heavy snowpack in the Merced River drainage started to melt rapidly, and the Merced flooded, backing the flow up the San Joaquin and inundating our weir and trap. With no barrier to block their way, the spring run moved to the dry reaches of the river, where spears and pitchforks ended their journey. A frustrated Fish and Game crew echoed the feelings of Fresno sportsmen who had suggested in a full-page newspaper ad that the agency responsible for the dry river should be called the Bureau of Wrecklamation.

Most of the spring-run salmon transported to the Outside Canal in 1948 spent the summer in the deep pools immediately below Friant Dam. From the north bluffs they resembled an accumulation of cordwood, lined up side by side facing the current, remaining almost motionless. By late September the salmon became active and were obviously ready to spawn. Since Friant Dam denied them access to the upper river, they dropped downstream to utilize the gravel riffles in the ten-mile stretch between Friant and Old Lanes Bridge.

Spawning was successful. In early 1949, substantial numbers of downstream migrants showed up at Mendota, where experiments with fish screens continued. Juveniles passing Mendota Dam reached the sack dam on the flow the bureau released for the Arroyo Canal. Below the sack dam, however, the trickle of water soon disappeared in the sand, stranding salmon migrants more than one hundred miles from the sea. The tragic conclusion to the history of the 1948 spring run was that the only beneficiaries of our efforts to salvage a valuable resource were the raccoons, herons, and egrets.

With little hope of obtaining water for progeny of a 1949 spring run, we decided not to repeat the 1948 trapping and hauling. The only hopeful alternative plan was to attempt to force the San Joaquin run to migrate up the Merced River, accompanying that stream's small native spring run. We installed a webbing weir, similar to the previous year's barrier, across the San Joaquin at the mouth of the Merced.

Salmon arrived as expected, but they refused to enter the Merced. Despite very poor water quality they pushed and probed at the webbing, trying to get up their home stream. The small San Joaquin was mostly warm return irrigation water loaded with salts and chemicals. In contrast, the Merced flow was clear, purer, and much cooler. But it was not home stream water. A few of the persistent San Joaquin fish were dipnetted, tagged, and released three miles up the Merced. These fish immediately turned downstream, almost beating the truck back to the weir!

Maintaining a fish-tight weir without night illumination was a problem. Poachers would slip in and slash the webbing to allow salmon to move upstream to shallow water where they could be speared or caught by hand. This damage would only be discovered and mended during debris removal operations the next day. This experience and others taught us about the salmon's uncanny ability to find even the smallest hole to get through a barrier. Another threat to the integrity of the weir was a tremendous migration of one- to twenty-pound carp that suddenly appeared. Tens of thousands of carp pushing on the webbing nearly lifted the heavy chain and sandbag anchors off the bottom. Some relief (and more problems) occurred when the Fish and Game Commission authorized a local fisherman to harvest carp at the weir.

Throughout May and early June, salmon continued to mill around, ignoring the ideal Merced flow. Those finding holes in the webbing quickly perished a few miles upstream. Below the weir, many salmon were taken by anglers using the "Armstrong Method"—snagging. Other fishermen used chickenwire dipnets on the pretext of bumping shad. While such poaching could not be condoned, it was difficult to condemn the taking of salmon illegally when a federal agency could eliminate an entire run of thousands of fish without qualms—and with no penalty whatsoever.

Probably none of the San Joaquin run ever migrated up the Merced River. These fish were either captured or succumbed to high water temperatures. The Fish and Game crews tried, but failed, to save the 1949 run. The final effort to salvage the lone remaining year class of San Joaquin spring-run salmon was scheduled in 1950. We knew that if we failed, this unique renewable resource would be lost forever. We also knew that the problems which had plagued us in previous years were still there: the dry section of river still prevented adult salmon from reaching the holding and spawning area above Mendota, and downmigrant juveniles still could not reach the sea.

Early in 1950, a minor concession by the bureau caused a completely different approach to salvaging the spring run. The bureau's offer of a temporary, tiny, release of twenty-five cubic feet per second of water for fish life gave us the opportunity to do something besides building weirs. It is debatable whether this release resulted from pressure from sportsmen and the Division of Fish and Game or whether, because of a good water year, more water was available than was required by agriculture. Twenty-five cubic feet per second in the riverbed below the sack dam would merely dampen the sand, but it might prove useful if it could be routed around the dry area. Assuming we could save the adult salmon, there was still no guarantee of water for downstream juveniles, but perhaps we had a foot in the door.

The quickly developed plan proposed diverting the fish release at the sack dam into the Arroyo Canal and then spilling it into Salt Slough to augment the slough's small flow of irrigation drainwater. Salt Slough was a meandering, tule-filled drainage ditch with abundant populations of carp, frogs, beaver, and voracious mosquitoes. But it was tributary to the San Joaquin River below the dry section, and the Arroyo Canal crossed it in a large flume, making it feasible to discharge the fish flow at that location.

To implement this plan, a concrete diversion structure and control works for a fish ladder were constructed at the approach to the canal crossing. A fish ladder leading up from Salt Slough to the flow control device was then fabricated of rough lumber. It consisted of six pools and "jumps" anchored in place with timber pilings. Two webbing weirs were also erected. The one installed at Salt Slough was attached to supports for the canal flume and shunted fish into the ladder. A second was placed at the mouth of Mud Slough, a tributary to Salt Slough, to prevent salmon from straying into that drainage.

The ladder functioned perfectly. The depth of pools and vertical rise between pools were within acceptable limits. The maintenance of effective weirs was another matter, however. Keeping the webbing barriers clean and functioning, a most disagreeable task, involved wading chest deep in filthy water to remove rotting carp and decomposing vegetation. It was also necessary to patch holes cut by beavers who objected to barriers on their water highways.

It would be gratifying to report that salmon streamed up the ladder in large numbers. This did not happen. Migration was deterred by high water temperatures, low flows, and, I suspect, poor water quality. In retrospect, I shudder to think what the levels of dissolved oxygen and total dissolved solids must have been, or the selenium concentrations. Furthermore, poaching was rampant along Salt Slough. Because the tule channel was narrow and shallow, salmon made "bow waves" as they moved along—easy targets for spears and pitchforks.

Only thirty-six salmon were counted at the ladder, and many of these were so weak they could barely swim from one pool to the next. I claim the dubious distinction of having had a personal involvement with the last four San Joaquin spring-run salmon. Some could just make it to the ladder, and I had to use a dipnet to carry them, one at a time, up the bank to release them in the canal. Probably none of the thirty-six lived to spawn.

So ends the saga of the San Joaquin River spring salmon. The Central Valley Fish and Game crew did the best it could, but that wasn't enough. Bureau of Reclamation officials and agricultural interests could now chortle that the troublesome Division of Fish and Game and irate local sportsmen no longer had anything to complain about. You may ask, "Why hasn't the bureau mitigated the loss of the San Joaquin run?" I could add, "Why haven't they mitigated damage to fishery resources caused by Shasta Dam, the Tracy pumping plant, or other of their projects?" I know the answer, but I won't state it here because you might resent the implication that I am pointing my finger at you.

Epilogue

Forty years have passed since the last spring salmon run ascended the San Joaquin River. During this time the Bureau of Reclamation, operator of Friant Dam, and its water contractors have never acknowledged their responsibility for causing the extinction of this irreplaceable resource, nor have they offered to mitigate the loss. In fact, as I write this, the San Joaquin water users are trying desperately to renew their existing water contracts with the bureau without undertaking an essential environmental impact study.

In addition to the salmon debacle associated with Friant Dam, the Bureau of Reclamation continues to operate the Central Valley Project with little regard for anadromous fish. The huge losses of juvenile fish at the Tracy pumping plant, fish passage difficulties at the Red Bluff Diversion Dam, and unfavorable water temperatures in releases from Shasta Dam are some of the problems created by CVP. As a climax to these unmitigated conditions, the bureau now claims to have found 1.5 million acre-feet of unallocated water in the CVP system that it wants to sell. (Delta outflow and other fishery needs be damned.)

On the brighter side, we no longer find the Department of Fish and Game and a few local sportsmen fighting and losing a rearguard action to save anadromous fishery resources. In the 1940s and 1950s I don't remember ever hearing anyone labeled an "environmentalist." We certainly have them today. Whatever you call them—nature lovers, protectionists, conservationists, sportsmen, commercial fishermen, or even bird-watchers—they are becoming well organized and their influence is being felt at state and federal levels and in the courts. The day has passed when disasters like the operation of Friant Dam can go virtually unchallenged.

Part Two

Current Perspectives

Salmon and steelhead have broad appeal. In this part of the book, each author addresses the subject in terms of his or her special interest. The tone and content of the selections, ranging from the abstract to the personal, illustrate the unique place salmon and steelhead occupy in our experience.

Patrick Higgins, a fishery biologist, opens this part with an explanation of how genetic adaptations developed in California salmon and steelhead and why genetics must be an integral element in all restoration plans. He suggests how this may be accomplished to obtain long-range benefits despite high initial costs. In dealing with the Klamath River basin salmon and steelhead fishery, Higgins draws attention to problems with clear genetic implications that exacerbate the strife among the principal groups involved: the Indian gillnetters, the trollers and sportfishermen, and the federal regulatory agency (PFMC).

In Chapter 7, Robert Ziemer and Richard Hubbard draw the reader into the fishery-forestry issue in their thoughtful essay on the problems of articulating timber harvest and fishery protection plans. Although the problems they raise defy solution, and seem to get worse as one thinks about them, their conclusion is optimistic: the topic is receiving much current attention in academia.

In Chapter 8 the scene shifts to the Sacramento Valley. Richard Hallock, closely involved in the Red Bluff Diversion Dam (RBDD) fish protection and enhancement operations for many years as Fish and Game biologist, discusses the project's problems, efforts to resolve them, and the current status of RBDD fish facilities. When that low dam on the Sacramento River began operations during the middle 1960s, it was hailed as a promising development in the struggle between fishery and irrigation interests. Its complex, state-of-the-art fish protection and enhancement facilities were a showpiece complete with visitors' center in a parklike setting where one could study colorful billboard-size illustrations and explanations of how the system worked. Several contributors to this volume touch on problems that led to the distressing current status of the RBDD and the associated Tehama-Colusa Canal fish protection facilities: they don't work the way they were supposed to. Worse, the dam has gained a reputation as the major recent cause of declines of salmon and steelhead runs on the upper Sacramento River. In this discussion, Hallock makes clear the reasons for the demise of this once-promising project.

In Chapter 9, fishery biologists Cindy Deacon Williams and Jack Williams offer a detailed account of a significant current crisis directly related to Red Bluff Diversion Dam: the almost certain slide into extinction of the Sacramento River winter chinook salmon. This species originally spawned and reared in cold tributaries of the Sacramento above the site of Shasta Dam and have hung on precariously in the river between Red Bluff and Keswick. The California–Nevada chapter of the American Fisheries Society and other conservation organizations met resistance from several sources, including the government, as they attempted to extend the protection of state and federal endangered species acts to the winter run. In 1989, the state Fish and Game Commission declared the species endangered and federal officials listed it as threatened under emergency regulations effective for two hundred and forty days.

"Salmon mystique," the special feelings we associate with salmon, becomes apparent in the next three chapters. In Chapter 10, economist Philip A. Meyer describes an economic approach to evaluating commercial salmon catches that takes into account a hitherto elusive human element: fishers' strong attachment to their life-style. With an economist's precision and charts, he challenges unrealistic assumptions that commercial salmon fishermen could as readily earn their living fishing in other places for other species.

In Chapter 11, Dave Vogel, whose fieldwork often extends into underwater explorations of salmon and steelhead habitat, relates a number of his sometimes harrowing experiences in a series of personal vignettes. He and his fellow scientists of the U.S. Fish and Wildlife Service extend their understanding of stream conditions for Sacramento River chinook and steelhead, and add to their scientific skills, in sometimes unorthodox ways.

In Chapter 12, Harvard sociologist Mary-Jo DelVecchio Good provides insights into the lives of a group we do not ordinarily associate with commercial fishing: women who fish independently and the wives of North Coast fishermen. Ashore, fishermen's wives worry about their husbands' safety and resent the disruptions of their home life. They function as important business partners. At sea, women who fish experience a primordial love of the ocean that they express as men rarely do. And when they learn the joys of catching fish, they become very competitive indeed.

The complex and emotionally loaded subject of Indian fishing rights is the focus of Chapter 13. Ronnie Pierce builds on the historical sketch she introduced earlier in Chapter 2 and explains her view of the current dilemma facing Indian and non-Indian fishers in the Klamath inriver and offshore fisheries. As a marine biologist and Native American activist, she leads the reader through the bewildering mix of issues that impinge on interminable efforts—"many tables" of meetings—to set appropriate allocations of Klamath River salmon among various user groups.

In Chapter 14 Bill Matson, a commercial troller who lives in the seaside hamlet of Trinidad, tells of his love of the sea and fishing for salmon. He also tells of problems that even the rugged individualists who are salmon fishermen feel powerless to overcome. The California North Coast troll fishery suffers from harvest restrictions imposed by federal management plans to protect wild stocks of Klamath River salmon. Ironically, the controls imposed on trollers resulted from passage of federal legislation, the Magnuson Act, which trollers backed because it promised protection from foreign competition. Now the competition is much more difficult to deal with because it is internal: a clash essentially between different cultures. Matson's advice: Quit fighting about who should get more fish, and concentrate on ways to improve the fishery for everyone, and the salmon as well. This chapter and those by Ronnie Pierce and Patrick Higgins constitute an introduction to the extreme complexities of the Klamath Management Zone fishery.

Joel Hedgpeth and Nancy Reichard make their special interest clear in Chapter 15: "Rivers Do Not 'Waste' to the Sea!" The general public tends to believe that streams expire as they find their way to the ocean. Wild young mountain streams become weary rivers, then doddering estuaries. Floods are particularly wasteful, because they destroy products of human effort and life itself. Such beliefs offer ready justification for major water projects. The slogan "Water must not waste to the sea!" was trumpeted in the 1950s campaign for the State Water Project and is still heard today. In this pointed essay, the authors argue convincingly that streams do not in fact waste to the sea—they have many beneficial effects as they find their way to the sea, and the public must understand that significant reality.

Part Two closes with "Steelie," the engagingly personal account of an outdoor writer's quest for the spawning grounds of wild steelhead high in a headwaters stream. Paul McHugh, who wrote this essay as a contribution to fishery restoration, tells more than an adventure story. He presents two other elements: a streamside glimpse at the basic biology of steelhead and a testimonial to the human fascination they engender.

Chapter Six

Why All the Fuss About Preserving Wild Stocks of Salmon and Steelhead?

Patrick Higgins

California's wild salmon and steelhead populations have an uphill battle for survival. Habitat for these fishes shrinks yearly due to a number of factors, such as logging and water development for agricultural irrigation and domestic use to accommodate California's burgeoning human population. To solve problems for the fish, many would argue that we should supplant wild fish production with hatcheries. Many fishery professionals disagree: that is not a wise course of action, and California has better options. My purpose here is to show why this is so.

Natural Selection

To the untrained eye, it may be very difficult to distinguish between a wild fish and one raised in a hatchery. The genetic information within each wild salmon or steelhead and those reared in a hatchery may vary considerably, however, and those genetic differences may have a profound effect on the long-term viability of California's salmon and steelhead runs and the costs to society to maintain them.

California's salmon and steelhead populations have a wide range of behavioral and physical characteristics that are controlled by the genetic code within each fish. That code has been molded by the success or failure of thousands of generations of these fish interacting with the physical and biological conditions of California's environment.

Behavioral traits allow salmon and steelhead to survive despite long-term trends toward a drier climate and seasonal droughts throughout much of the state. If a run of steelhead spawns in the upper reach of a stream that is near a spring, for example, and the lower stretches of the creek dry up in summer, the offspring may have genes that tell them to stay near the spring. Also responding to genetic signals, the fry of earlier-spawning chinook salmon in the same stream might migrate downstream quickly after emerging from the gravel to avoid this problem.

Variations of such patterns may also be observed. For example, runs of coho salmon and steelhead adapted to small coastal streams near Santa Cruz are flexible in the time of their return to fresh water. If rains come early, some of these fish may return in October. Should drought conditions exist, they may not spawn until January. Indeed, some southern California coastal streams host steelhead runs that return only in years of abundant rainfall.

One of California's last substantial runs of summer steelhead on the Middle Fork of the Eel serves as a classic example of genetically controlled survival strategies. As the last snow melts from the Yolla Bolly Mountains, these fish battle up the steep, rocky Middle Fork gorge. Streamflows drop rapidly, but the summer steelhead seek shelter in deep pools that are cold on the bottom. Here they wait, often as if in suspended animation, until fall rains allow them to spawn in tributaries. The offspring of all the steelhead in this drainage are also able to thrive in summer water temperatures that are consistently above seventy degrees. Less hardy strains of steelhead fry would be lethargic and susceptible to disease in these elevated water temperatures.

Many of California's rivers once had numerous distinct runs of salmon and steelhead returning throughout the year. The evolution of variations in timing between runs in the same river is due at least in part to biological interactions changing genetic makeup. By entering the river and spawning at different times, competition for spawning and rearing areas is minimized. As fry of the different runs emerge from the gravel, they take turns using different parts

Natural home. This small tributary of the Smith River in northern coastal California illustrates ideal nursery habitat for juvenile salmonids: forest canopy and streamside vegetation, oxygenated flow, deep pools, and cobbles and gravel streambed. (California Trout, Inc.)

of the stream habitat. Chinook salmon may be the first to return to spawn in the fall, followed by coho salmon, and finally steelhead in mid to late winter. As the chinook salmon emerge, they feed in the slow water at the edges of the stream. By the time the coho salmon fry emerge, the chinook salmon are feeding in swifter, deeper waters and beginning to move downstream. Thus competition for food and space is minimized.

California's stream systems are endowed with salmon and steelhead specifically adapted to the geology, hydrology, and ecology peculiar to each. Many of these native strains have been lost where habitat was completely destroyed, but others have survived against amazing odds. For example, suburban development in Pacifica, south of San Francisco, caused a whole tributary of San Pedro Creek to be placed in a culvert. Years afterward, steelhead in that tributary were still spawning far up that dark culvert in the gravels deposited in it. Where native strains remain they must be protected. The preservation of even remnant runs could play an important role in rebuilding the state's salmon and steelhead stocks.

Restoring Salmon and Steelhead Runs

In recent years considerable efforts have been made to restore habitat and salmon and steelhead runs in California rivers. Hillslopes have been stabilized, streamside vegetation replanted, and structures to improve habitat placed in streams. Runs of salmon and steelhead in these watersheds are typically depressed when these restoration projects are initiated.

Some of the few returning adult fish are trapped and spawned artificially and eggs raised in "hatchboxes" to increase survival rates. These native fish survive well in their restored habitat and begin to reproduce naturally. As we continue efforts to restore California's fishery habitat, we may begin to encounter areas where no remnants of native strains exist. The more wild populations of salmon and steelhead that remain throughout the state, the better the chances are that the appropriate genetic strains for natural reproduction in the widest range of environments will have been retained.

Genetic engineering is much in the news these days, and one might get the impression that if we needed a fish with certain characteristics, we could just mold the DNA of a salmon or steel-head. Actually, we are only now beginning to decipher the genetic code of these fishes. We know that there are more than a thousand different points of information, or loci, on the gene. Of these loci, we understand the function of only a dozen or so. Physical or behavioral traits can also result from interplay between combinations of genetic sites—combinations that are almost infinite in number. In short, these fish have evolved adaptive mechanisms that we do not understand. We cannot genetically engineer traits for survival, therefore, unless we retain these fish and copy from their genomes. Although salmon and steelhead sperm can be preserved by freezing, there is no known way to preserve the eggs of these fishes.

If we lose this precious genetic resource, our watersheds may remain perpetually underseeded or we will have to rely more and more on artificial rearing to maintain poorly adapted populations. Costs will escalate to maintain salmon and steelhead in the state, valuable genetic resources for future aquaculture or enhancement efforts will be lost, and an important part of our heritage will have disappeared as well.

Problems with Hatchery Production

California's first hatchery was opened in 1872 on the McCloud River, a tributary of the Sacramento. Despite this early start on artificial production and a tremendous investment in hatchery facilities associated with California water projects, hatchery success in rebuilding or maintaining stocks has been a hit-and-miss proposition. Problems with disease or human error in operation can cause catastrophic fish kills. Heavy dependence on hatchery-produced salmon and steelhead is a very expensive solution to diminishing runs of these fish and one fraught with problems.

Typically, hatcheries in the past selected the first fish returning to the hatchery for breeding purposes. Hatchery managers wanted to capture enough adult fish at the earliest opportunity to acquire sufficient numbers of eggs to utilize fully the incubation and rearing capacity of their facility. We now realize that this method may have saved only a fraction of the genetic information and survival strategies that existed in the genes of a broader cross section of the salmon or steelhead in that drainage. Apart from the obvious effect of causing runs in hatchery-dependent rivers to be early and of short duration, various life-history patterns and survival strategies are lost when genetic diversity is restricted in this fashion. Ocean migration and other patterns are often virtually uniform in hatchery stocks. As ocean conditions or streamflows vary from year to year, the diverse strategies of wild fish can dramatically increase their chances for survival over hatchery fish. Returns of salmon and steelhead that are genetically restricted are subject to much wider fluctuations in rates of return. By retaining stock diversity, populations of these fishes will be more stable.

If genetic diversity becomes too restricted, reproductive capacity declines. This phenomenon, called inbreeding depression, results in decreasing hatchery production, which may ultimately necessitate replacement of the entire hatchery broodstock. A disease outbreak can be devastating in a hatchery if the strain being raised lacks genetic resistance to the pathogen. The Coleman Hatchery on Battle Creek, a Sacramento River tributary, has had continuing problems with outbreaks of disease. In worst-case scenarios, whole broodstocks from hatcheries may need to be replaced because of disease.

Human error can result in tremendous fish kills, which can greatly alter returns in hatchery-dependent rivers. In the summer of 1986, Iron Gate Hatchery on the upper Klamath River experienced a fish kill of one to three million juvenile chinook salmon being reared in cold hatchery water. As repairs were being made to the dam upstream, water was being released from the surface instead of from the bottom of the lake behind the dam. Water temperatures below the dam were in the high seventies, and when the young fish were released almost all of them died.

But hatcheries are not all bad. Despite the problems inherent in hatchery production, hatcheries play a vital role in supplanting losses where habitat has been irretrievably lost. In the 1960s, new feeding methods and ways of combating disease were discovered, and production levels at hatcheries generally increased. Recent investments by commercial fishermen on improvements to Sacramento and San Joaquin River fish-rearing facilities through the Salmon Stamp Program have also resulted in substantial increases in hatchery production in the state. As hatchery production increases, negative side effects on native strains can occur, so special care must be taken by hatcheries in basins where viable wild populations still exist.

The Impact of Hatchery Fish
on Wild Populations

Fishery professionals used to believe that hatchery programs which produced numbers of fish above the numbers of naturally produced offspring had little or no negative impact on wild runs. We are now, however, discovering that hatchery programs and the way in which hatchery fish are transferred and planted can have substantial negative effects on wild fish.

California has opted for large hatchery facilities on just a few rivers as opposed to smaller facilities on a greater number of streams. On the surface, the strategy is easy enough to understand: large hatcheries have larger production capabilities and therefore lower cost per fish produced. Also, most of these facilities were funded in association with dams for mitigation. When these large hatcheries have surpluses of juvenile fish, the fish are often transferred to "enhance" populations in other streams. The fish may be poorly adapted to the watershed to which they are introduced. They are also sometimes planted without regard for the carrying capacity of the stream. Competition for food and space can greatly reduce the number of native juveniles in these systems.

Hatchery fish often do not survive as well in the stream or ocean as wild fish, yet as thousands or millions of fry or smolts are planted in streams, they outcompete native populations because of sheer numbers. Hatchery fish are taught to respond positively to humans or machines moving along raceways to feed them. In the wild, such movement over the water might be a predatory bird or animal. Wild fish stake out territories where little energy is required to stay in feeding position but much food is delivered. Hatchery fish may not be imprinted to behave in this fashion. We are just now gaining greater understanding about timing the release of these young fish. If they are not released to begin migration when genetic cues normally would induce such behavior, they may not even go to the ocean at all.

Transferred stocks are often poorly adapted to conditions in host streams. Coho salmon from hatcheries on large northstate rivers have been planted in short coastal streams near Santa Cruz. The genetic information they contain gears them for a journey of several hundred miles, but the systems to which they have been transferred are only a few miles long. Coho salmon returning to the Klamath, for instance, never encounter low streamflows near the mouth and begin their upstream migration several months before spawning. When these fish are transplanted, they might return to spawn when the Santa Cruz area streams are almost dry. Native stocks of California coho salmon have declined seriously, partly because of such stock transfers.

Straying of returning adult spawners is also a problem with hatchery fish. Instead of returning to the hatchery, they may continue upstream, enter a tributary below the hatchery, or return to a different river system. This problem is compounded if the salmon or steelhead are planted away from the hatchery facility, as in the estuary of the river or in an entirely different stream.

Hatcheries often plant juvenile salmon in estuaries because the number that survive to maturity is greatly increased. Intermixing of hatchery and wild stocks may, however, introduce genes into the wild population that decrease survival. A prime example of this is when introduced fish lack resistance to a disease organism present in their new environment. The result can be catastrophic.

Wild fish can also be severely affected by overfishing when harvested in mixed-stock fisheries with hatchery fish. Hatchery stocks can sustain harvests of up to 90 percent, while wild populations may be jeopardized by any harvest pressure over 65 percent. The reason for the difference is the higher survival rate of eggs reared in hatchery trays over eggs that are exposed to the many hazards of the natural environment. To maintain maximum genetic diversity and protect wild runs, harvest quotas must be set to preserve the more vulnerable native fish.

If harvests are kept at a rate to sustain wild populations, thousands or even tens of thousands of surplus fish may return to the hatchery gate. Huge concentrations of fish in the stretches of river below these hatcheries overwhelm the river's spawning and rearing capacity. Many fish that are not allowed in the hatchery die without reproducing. These massive runs encourage poachers, who rationalize their "sport" by saying that the fish will otherwise be wasted.

Legitimate sportfishing in rivers supporting these huge hatchery runs may also reach levels that deplete native runs.

Hatchery returns may also mask problems with native stocks and create false impressions of the overall health of the salmon and steelhead runs in certain basins. While runs to the hatcheries in the Klamath/Trinity River system have boomed since 1984, wild fish in drainages like the South Fork of the Trinity have not rebounded at all. Wild stocks were overharvested for so long that rebuilding will take many years. Habitat problems may also be besetting these native fish.

Study of the decline in populations of older forty- to seventy-pound Klamath River salmon has revealed a perplexing facet of this problem. Commercial and sportfishing pressure over the years has reduced the numbers of four- or five-year-old fish to the extent that they have nearly disappeared from the Klamath and other California streams. Spawning females of these stocks produce up to ten thousand eggs. Such fish may be genetically adapted to spawning deep in large rivers, using habitat where smaller salmon are unable to reproduce successfully. Without fish to utilize it, this habitat may not be used for spawning. As ocean fishing regulations have tightened in recent years, more of these large fish seem to be returning.

But another set of problems must be dealt with. The large-mesh gillnets used in the Indian fishery on the Klamath River selectively harvest the largest returning salmon, thus thwarting efforts to restore these important stocks. The large mesh does permit the escape of smaller steelhead, however, an important sport fish in the Klamath basin. A shift to smaller nets would, therefore, not only reduce the value of the Indian salmon catch but also produce negative side effects on the important sport fishery, which contributes significantly to the local economy. Solutions to problems of preserving diverse runs are never simple.

How to Protect Wild Stocks

To protect California's remaining wild salmon and steelhead we must, first and foremost, fight to protect remaining habitat. As we work to restore California's salmon and steelhead resource, we are finding the task oftentimes difficult and costly, so preservation of existing habitat makes economic good sense as well.

In areas like the North Coast, where considerable habitat remains that can support wild salmon and steelhead, care must be taken to lessen the negative effects of hatchery fish on wild fish. The original fish used as parents or broodstock at the hatchery should be taken from a wide spectrum of the native run. New wild fish should be captured periodically and used in breeding. In this way the fitness of the hatchery strain is maintained, and if some straying occurs, negative impacts are minimized. If a stream's entire run of salmon or steelhead is extinct, broodstock should be chosen from the nearest similar basin so that natural production has a greater chance of being reestablished. Selective breeding at the hatchery for large size (or any other trait) should be avoided. While individual fish in returning runs might tend to be larger, many traits that would contribute to long-term survival might be lost.

California hatchery managers need to pay more attention to the carrying capacity of the stream or river where juveniles are released. In years when tremendous numbers of progeny are reared, survival rates may be very low after release because the natural system is incapable of supporting the young fish. Survival of wild fish under these crowded circumstances might also be very low.

Salmon or steelhead juveniles generally should not be planted away from their hatchery of origin. Young fish released at the hatchery are more properly imprinted and therefore stray much less to spawn with wild fish. Transferring juvenile salmonids between basins should be avoided under most circumstances.

Hatchery juveniles are often much larger than wild salmon and steelhead young at the time of their release. Hatchery programs should release fish equal in size to native fish in the same habitat so they will not prey upon wild young or have an unfair competitive advantage in seeking food or habitat.

Hatchery fish should be harvested more intensively than wild fish. If hatchery fish were all marked, they could be selectively harvested in the river and in the ocean, and wild fish released. British Columbia, Washington, and Idaho currently manage their steelhead in this manner. British Columbia's anglers initially grumbled at not being permitted to keep unmarked steelhead, but now fishing has improved so dramatically that they are extremely happy with this management strategy. Managing salmon in this fashion is more complex logistically, but it may be a worthwhile strategy to explore.

The California Department of Fish and Game is caught in a bind on the issue of protecting wild stocks. As the state's population increases, problems of fishing pressure also increase. Many decision makers within the department see the only solution as increasing artificial production of salmon and steelhead. In certain cases, like the San Joaquin River, where the native fish have been wiped out, hatcheries may be the only solution. But much of the state's salmonid habitat remains viable, and more effort is needed to maintain and enhance wild salmon and steelhead populations by employing management strategies to put harvest in balance with production wherever possible.

In the long run, preservation of wild fish makes economic good sense. When wild fish come back to spawn, it doesn't cost California's taxpayers a dime.

Chapter Seven

Forestry and Anadromous Fish
Robert R. Ziemer and Richard L. Hubbard

One of the most pressing . . . problem[s] involves the effect of timber harvest upon fish resources. . . . Rarely has so much discussion been generated around so few facts.

—D. W. Chapman, 1962

This statement is nearly as applicable today as it was when made more than twenty-five years ago in the *Journal of Forestry.* The relationship between forest practices and anadromous fish production has continued to be debated during the intervening decades without a clear resolution. The issue is complicated because there are activities in addition to forest practices that affect anadromous fish production. The offshore fishery removes a large portion of those adults that would return to the streams to spawn. Instream fishing removes another portion of those spawners. Dams on the rivers reduce peak streamflows that influence channel morphology and sediment transport, as well as modify low-flow discharges in the summer. Much of the downstream river habitat is modified by major highways, agriculture, and urbanization. Estuarine habitat has been virtually eliminated from many rivers and severely modified for the remainder.

In the forested areas, past and present land use is variable. Many mountainous watersheds were severely modified by extensive placer mining and logging during the last century. In the late 1940s, the increased value of softwood species, such as pines and firs, started a new wave of cutting in the forests. Beginning in the mid-1950s, large storms reactivated huge dormant streamside landslides. Not until the 1970s did forest practices legislation begin to address issues of riparian condition and habitat. The question facing researchers now is how to separate all of these influences, including the effect of past forest practices from present and future practices in relation to fish production. The important regulatory challenge is to be able to predict the influence of new activity given the present condition of the resource of concern.

In 1987, at the request of the California Advisory Committee on Salmon and Steelhead Trout, the Wildland Resources Center of the University of California convened a workshop at the U.C. Davis campus to define the needs and costs of a ten-year research, development, and education program related to salmon and steelhead trout. A cross section of commercial and sportfishermen, government resource managers, university scientists, and consultants compiled a list of one hundred and thirty-nine problems needing solution. From that list, eighteen problems were given highest priority for expanded funding and research. Two of those eighteen problems are directly related to the forestry and fishery interaction:

1. Determine how changes in inputs of sediment and associated changes in instream channels affect fish habitats under varying conditions.
2. Identify and assess the cumulative effects of timber harvest on erosion, hillslope stability, streamflow, and sediment in stream channels.

After decades of work, we still cannot predict biotic changes from measured changes in the physical environment of watersheds or stream channels. This limitation has, in some cases, resulted in the destruction of habitat in the name of protection. Until recently, forest practice regulations addressed water quality—not fish habitat. Our view of woody debris, for example, has changed dramatically over the past decade. Programs to protect water quality at times required extreme measures to clean up streams after logging. Occasionally these programs were translated into removing all woody debris from the stream—both natural and logging-induced. Often the result was accelerated erosion of channel beds and streambanks. Large woody debris is now recognized as an important component of healthy streams. It moderates the velocity of streamflow, influences the routing and storage of sediment, and increases the quality and diversity of fish habitat.

Most forest practices regulations, and most research on land use effects, have focused on short-term responses of local areas to single land uses. These responses are typically viewed as being isolated in time and space. Recently managers and researchers (and the courts) have become increasingly concerned with the "cumulative effects" of land management activities. The National Environmental Policy Act defines such effects in this way: " 'Cumulative impact' is the impact on the environment which results from the incremental impact of the action when added to other past, present, and reasonably foreseeable future actions regardless of what agency or person undertakes such other actions. Cumulative impacts can result from individually minor but collectively significant actions taking place over a period of time."

Cumulative environmental changes may occur either at sites of land use disturbance or away from the disturbed sites. At the site of disturbance, multiple practices may combine or accumulate through time to affect a beneficial use. Away from the site, changes may accumulate through a sequence of impacts spread over many years or through the combined effects of multiple practices distributed throughout the river basin. Concern about cumulative effects introduces the concept that even though all activities are conducted in a manner which limits their individual effects to an acceptably low level, unacceptable harm may be experienced at some point in time or space when these activities function collectively.

Today we have no effective method for predicting the environmental response to a land use plan. To make matters worse, there is little agreement among disciplines, geographic regions, or interest groups over what actually constitutes cumulative effects or whether they even exist.

Determining the influence of land use on resident fish, let alone anadromous fish, is particularly problematic. First we need to understand how land use activities affect the removal over time of sediment, water, woody debris, nutrients, and heat from hillslopes and their delivery to streams. Then we need to know the transport rates of each of these products from the sites of land use to areas of concern. Moreover, we must determine how altered sediment, water, woody debris, nutrient, and heat transport affect resources of concern, such as diversity, composition, resilience, and structure of biological communities. Finally, since fish are near the top of the biological community structure, we need to understand the importance of these changes on not only the fish but also their ecological link to other parts of the community throughout their life cycle. It is much simpler to understand, for example, how a single land use activity over a short period affects erosion than to understand the biological consequences of the resulting sediment.

There are important issues related to scale—both spatial and temporal. In general, individual erosion events are limited to an area of square yards or, at most, acres. Individual land management activities, such as logging, usually occupy less than a hundred acres. The drainage area of the streams that contain most of the prime anadromous fish habitat exceeds a thousand acres, and usually more than ten thousand acres. In small areas, it is relatively simple to measure the relevant variables in order to evaluate cause and effect. As the area becomes larger, it becomes progressively more difficult to measure these variables at a scale that can give meaningful results. And as the spatial scale increases, so does the time required for a change to be observed. For example, the time required for sediment to be routed from a site of erosion within a one-acre watershed is much less than in a hundred-acre watershed or a ten-thousand-acre watershed. Therefore, the relevant response time between a land use activity and a significant effect should be expected to increase as the size of the area increases.

Similarly, the recovery time following disturbance should be expected to increase as drainage area increases. As the time between disturbance and expected effect increases, there is a greater chance that a natural event, such as a major storm, will occur within that interval and confuse any determination of cause and effect. In some cases, land management decisions can set the stage for substantial problems during serious storms. If conventional road engineering designs call for forest road culverts to withstand a fifty-year storm, for example, then during a hundred-year logging cycle all of the culverts would be expected to fail twice, on the average, and the associated road fill would be washed into the stream. This is not a natural consequence of a large storm; it is an economic and design decision.

Even when we eventually understand the relationship between land use practices and erosion and sediment production and routing, we will still be a long way from understanding the effect of that sediment on the biological community, including anadromous fish. The important point to keep in mind is that none of these relationships are simple. To evaluate the effect of logging on sediment production immediately below the area of activity is not enough when the area of interest is ten miles downstream. Furthermore, to understand the effect of that logging operation on the sediment regime ten miles downstream is not enough when the objective is to understand the effect of sediment on anadromous fish production.

It now becomes important to know the change in flux of that sediment throughout the life cycle of the fish and the effect of these changes upon growth, reproduction, and mortality. The effect may not be direct, but it may represent a change in food availability, feeding success, susceptibility to disease, or predation. Thus we must be concerned not only with the immediate effect of the sediment on the fish but also its effect on the ability of the fish to grow, compete, and eventually reproduce. If a change in sediment load, for instance, lessens the ability of a fish to survive and reproduce, that is perhaps as important an effect as killing the fish outright.

For several decades, the riparian zone has continued to be the focus of increasingly restrictive regulations—and for good reason. Thirty years ago the riparian zone was a place to locate roads, landings, and skid trails. Logs were routinely tractor-yarded to and down stream channels. Large volumes of soil and logging slash were left in the streams. Road construction debris was routinely side-cast, much of it in the stream. Studies of land management effects on fish usually focus on stream blockage by logging debris and lethal temperature increases resulting from removal of the tree canopy. More recent fish management programs have called attention to additional specific habitat requirements, such as spawning substrate, sedimentation, cover, pool volume, minimum instream flows, and the effect of land management practices on these requirements. Single-objective programs—for example, to increase the amount of suitable spawning substrate—often do so in the absence of the necessary collateral knowledge of sediment transport mechanics, channel morphology, and other aspects of fish habitat. Such programs are often a disappointment; they do not attain the objective of increased return of adult salmonids. The programs fail because they ignore major attributes of stream ecosystems that support the fisheries.

Clearly an ecosystem approach at an appropriate spatial and temporal scale will be required if progress is to be made on the question of forest practices and fish production. Regulating individual timber harvest units is not enough. Regulation must be made at the drainage basin scale, taking into account the effects of past and present practices. It is not sufficient to have streamside management regulations designed to maintain stream temperature. Regulations must also consider changes in streamside input of solar radiation, nutrients, food, litter, woody debris, and sediment over both the short and the long term.

As one example, management decisions in the riparian zone can substantially affect the supply of large woody debris without significantly affecting the other streamside inputs. If the management policy is to harvest continually only the large and decadent trees, leaving the vigorous intermediate and small-sized trees, little change would occur in any of these other streamside additions. The incidence of tree-fall, however, would be dramatically lowered. The quantity of large woody debris in the stream would gradually be reduced by stream export and decay, but new additions of large material would seldom be available. Eventually the stream would become devoid of large woody structures and the morphology of the channel would adjust, as would cover and other aspects of the aquatic habitat that are tied to the presence of large pieces of wood.

Besides transporting water, the stream transports sediment, nutrients, detritus, and organic matter from the surrounding forests and hillslopes. The riparian zone links hillslopes to streams and moderates the transport and delivery of these watershed products. The riparian ecosystem functions within the context of changing fluxes of these products, and anadromous fish use the streams draining the forested watersheds for only a portion of their life cycle. A recent symposium at the University of Washington assessed the state of the science on forestry/fish interactions. A reading of the 471-page proceedings clearly demonstrates that there has been progress in understanding pieces of the forestry and fish puzzle,

Tractor yarding equipment at a logging site. More stringent rules, coupled with stricter enforcement, are needed to reduce damage to streams from such operations, particularly in northern coastal California watersheds. (Herbert Joseph)

but much work remains before we can predict the effect of a proposed land treatment on fish production in any given drainage basin.

Since D. W. Chapman discussed the issues of forestry and fish resources a quarter-century ago, the populations of salmon and steelhead trout have continued to decline. Robert Z. Callaham, of the Wildlands Resources Center, U.C. Berkeley, and Bruce Vondracek, Department of Wildlife and Fisheries Biology, U.C. Davis, point out that "reversing the decline depends, in part, upon having new technology to improve management of these fisheries and that technology would be applied by a strong research, development, and extension (RD&E) program." The needs and costs of an RD&E program to improve the management of salmon and steelhead trout have been identified by others. Because of the importance of salmon and steelhead trout resources to the economy of the Pacific Northwest, these programs, recommended by commercial and sportfishermen, government resource managers, university scientists, and consultants, need financial and political support to move beyond the planning stage to implementation. The California Advisory Committee on Salmon and Steelhead Trout, in its 1988 report, emphasizes the urgency of the task: "California must aggressively confront the problems challenging salmon and steelhead survival. It is not too late to restore and protect this natural heritage. The time to act is now."

While complete reversal of anadromous fishery declines will depend on results of the research described above, promising interim actions are being taken. Tightening and better enforcement of the State Forest Practice Act is one such action. The high priority given to fisheries by the U.S. Forest Service, as outlined in its "Rise for the Future" program, is another such action. The Bureau of Land Management has announced that it intends to address fishery problems more vigorously. These actions, coupled with an ambitious research and development program, are certainly a glimmer of light at the end of what has been a very dark tunnel.

Chapter Eight

The Red Bluff Diversion Dam
Richard J. Hallock

One of the major causes, and perhaps the single most important recent known cause of the decline of salmon and steelhead in the Sacramento River is Red Bluff Diversion Dam (RBDD). Completed in 1964, the U.S. Bureau of Reclamation's diversion dam is located on the Sacramento River two miles downstream from Red Bluff. Initially incorporated into the diversion complex were state-of-the-art fish protection facilities that were plagued with problems and modified structurally and operationally. Key features were determined to be unusable in recent years.

The bureau presently maintains and operates RBDD and the upper two or three miles of the Tehama-Colusa and Corning canals, including the Corning Canal pumping plant. The Tehama-Colusa Canal Authority, an association of water users, now operates and maintains the remainder of the Tehama-Colusa and Corning canals. The Tehama-Colusa Fish Facilities have been deactivated, and the salmon spawning and rearing areas are no longer in use. Some of the funds formerly allocated to operation of these facilities have been transferred to Coleman National Fish Hatchery; a portion is still used for maintenance of the deactivated facilities. The Tehama-Colusa Fish Facilities offices, shops, and storage buildings are now occupied by the U.S. Fish and Wildlife Service's (USFWS) Fisher-

This essay is condensed from *Sacramento River System Salmon and Steelhead Problems and Enhancement Opportunities*, a report prepared by Hallock for the California Advisory Committee on Salmon and Steelhead Trout, June 1987; updated June 1989.

Sacramento River chinook salmon like this one typically weigh from twenty to thirty pounds. (California Department of Fish and Game)

ies Assistance office. This office, among its other duties, operates and maintains in part the fishways at RBDD and is also responsible for fish counting operations there.

The Bureau of Reclamation is presently replacing the inefficient louver-type fish screen at the entrance to the Tehama-Colusa Canal

Part of a grand design that failed. Despite hopes of engineers and fishery
planners in 1966, the Red Bluff Diversion Dam is now considered a major
cause of recent declines of salmon and steelhead populations in the upper
Sacramento River. (Dave Vogel)

with a positive, 32-revolving-drum type of fish screen. The bureau
also is increasing canal diversion capabilities from 2,800 to 3,200
cubic feet per second by adding one new bay at the entrance.
During an average water year (without the new bay) 700,000 acre-
feet of water is diverted into the Tehama-Colusa Canal and an
additional 50,000 acre-feet is diverted into the Corning Canal. The
diversion headworks near the right-bank abutment is screened to
prevent fish in the river from entering the canals.

Sacramento River water levels are controlled by eleven dam
gates, each 60 feet wide and 18 feet high, incorporated in the 752-
foot-long structure. Water is released by raising one or more gates.
A fishway, with closed-circuit television to count adult salmon and
steelhead, is located on each dam abutment. A fish trap is incorporated into the left-bank fishway, where adult salmon and steelhead
can be examined and released or selected for transfer to other
locations.

Fyke net being repaired. These nets are used for sampling fish populations for study. (California Department of Fish and Game)

Spawning Distribution Changes

Shortly after RBDD became fully operational in 1966, portentous changes occurred in the distribution and total number of fall-run salmon using the upper Sacramento River system. Prior to 1966 more than 90 percent of the total salmon population spawned above the dam site and less than 10 percent below. During the first decade of the dam's operation, the spawning distribution pattern changed: less than 40 percent were spawning above the dam and more than 60 percent below. Changes in these percentages have coincided with overall declines in total populations. Thus both relative and actual numbers of salmon that spawn in waters above the dam have dropped dramatically.

Problems

The problems at RBDD are primarily related to passage of both adult and juvenile salmonids. Adult salmon are delayed below the dam from one to forty days, and more than 26 percent that approach the dam never get past it. Of the four salmon runs, adult fish passage problems cause the most damage to winter-run salmon. Delay time, which adversely affects spawning success, increases with increases in flow, since the adult fish have more difficulty finding the fishways at higher flows. Juvenile salmonids that do not have to pass the dam on their way to the sea have a greater chance of survival than those that do—fingerling fall-run salmon 46 percent greater and yearling steelhead up to 25 percent greater. Studies conducted in 1974 by the U.S. Fish and Wildlife Service showed losses as high as 55 to 60 percent being suffered by juvenile salmon passing the dam during daylight hours.

Fish losses specific to RBDD are caused in part by (1) inadequate attraction flows from the fishways, which result in delay and blockage of adults moving upstream, and (2) turbulence immediately below the dam, which disorients both juvenile and adult salmonids. In particular, the juveniles moving downstream are thrown to the surface after passing under the dam gates, where they become easy prey for predatory fishes, especially Sacramento River squawfish. Other documented losses of juveniles have resulted from the canal headworks fish screen.

Fish Losses

Historical data are lacking for all but fall-run salmon, resulting in less accuracy in estimating the effect of RBDD on late fall-, winter-, and spring-run salmon, as well as steelhead. Between 1969 and 1982, however, RBDD has caused an estimated loss in the upper Sacramento River system's adult salmon populations of 114,000 fish: 57,000 fall run, 17,000 late fall run, and 40,000 winter run. These losses have deprived the fisheries of about 228,000 salmon a year at a catch-to-escapement ratio of two-to-one. Other researchers agree with such estimates. In a 1986 report to the USFWS, R. R. Reisenbichler estimated that solving the problems at RBDD would return the fall-run salmon population to 1955–1956 levels. In addition, an estimated decline of 6,000 sea-run steelhead in the upper Sacramento River has been attributed to RBDD.

Problems of Handling Ripe Salmon

The Department of Fish and Game routinely samples fish migrating upstream in the trapping facility at RBDD to separate the total closed-circuit counts into the various runs and to look for marked and tagged fish. Between 1971 and 1974, about fourteen hundred ripe female salmon (losing eggs when handled) with an estimated average potential of seven million eggs were handled annually. The number of ripe females handled currently (1989) would no doubt be less, even with increased sampling, because of population declines, especially in winter-, spring-, and late fall-run salmon. At present, these fish are still released in hopes that they will eventually spawn successfully, but success seems unlikely.

To solve this problem of ripe spawners, the U.S. Fish and Wildlife Service constructed an incubation station near the left-bank fishway that became operational in 1979. It has not been used to date, primarily because of lack of personnel and management interest. The handling of seven million eggs in this facility annually could have added between seven thousand and thirty thousand fish to the ocean catch, depending upon their size when released. Moreover, this procedure could have given a boost to the now endangered winter-run salmon, since in the 1971–1974 period more than a million of the total eggs would have come from ripe winter-run salmon in May and June. Since the winter-run chinook population declined to about five hundred fish in 1989, such an egg take would no longer be possible. The problem is further compounded by the fact that the incubation station will apparently never be used for its original purpose, since it is now being transferred by the USFWS to the U.S. Forest Service, managers of the adjacent recreation area.

Squawfish Predation

Between 1978 and 1985 the number of adult Sacramento squawfish counted annually as they passed upstream through the fishways at RBDD ranged from a low of thirteen thousand in 1983 to a high of twenty-five thousand in 1978 and averaged about eighteen thousand. Squawfish concentrate below RBDD in the spring and early summer, where they prey heavily on juvenile salmon and steelhead on their way to the sea. Turbulence caused by large volumes of water flowing under the dam gates disorients the juvenile salmon and increases their vulnerability to predation immediately below the dam. Squawfish sampled below the dam during two sampling periods in June 1977 had consumed an average of 0.5 and 1.5 juvenile salmon shortly before capture. In May and June 1977, an estimated twelve thousand squawfish were concentrated below RBDD—representing a potential daily consumption rate in excess of one hundred thousand juvenile salmon. During the spring and summer months of especially dry years, striped bass also become quite numerous and are serious predators of juvenile salmon immediately below RBDD. For example, during one study the stomach of a twenty-five-inch-long striped bass captured below the dam was found to contain the remains of twenty-one juvenile salmon. Studies in April and May 1984 showed that squawfish predation was causing losses among juvenile salmon as high as 55 percent during the daytime.

To control squawfish at RBDD an electronic shocking device was installed in the left-bank fishway and tested in 1985. This device was quite successful in destroying adult squawfish in the fishway as they were migrating upstream. Its operation had an adverse effect on salmon migration, however, so use of the shocker was discontinued. Apparently when squawfish, and certain other species, are under stress a warning odor is emitted. In 1987 a new device, its purchase funded in part by the Marin Rod and Gun Club, was tested in the left-bank fishway. Its purpose was to reduce stress by capturing squawfish alive in the fishway and then destroying them elsewhere.

Currently, electronic device testing is at best intermittent since the gates at RBDD are raised for a four-month period between December 1 and April 1 to provide "free" adult passage for the endangered winter-run chinooks. The National Marine Fisheries Service also funded a study to determine whether a commercial squawfish fishery might be feasible. A contract was let to a commercial fisherman to remove squawfish at RBDD, but harvested fish cannot be sold commercially because their flesh has been found to be contaminated with dioxin.

Lake Red Bluff Power Project

The city of Redding applied to the Federal Energy Regulatory Commission (FERC) in 1983 for a license to operate Lake Red Bluff Power Project. FERC denied the permit, but Redding has appealed. The city of Redding's plan is somewhat similar to a plan developed by the Bureau of Reclamation to develop power at RBDD—a plan that the bureau is not actively seeking to implement at this time.

A major concern with the city of Redding's proposed power project is the potential direct turbine mortality of juvenile salmon and steelhead migrating downstream—that is, those fish which cannot be diverted or screened from passing through the turbines. Indirect mortality—that is, increased predation on stunned, disoriented, or debilitated juveniles that have passed through the turbines—could also be significant. Adult salmon and steelhead passage upstream at RBDD could also be adversely affected, since the proposed project provides for inadequate fish attraction flows to the fishways.

Recommendations

To help solve adult fish passage problems at RBDD, both fishways should be modified to provide exit flows two to three times what they are now. At the same time, comparative evaluations should be made of proposals to improve fish passage by further enlarging the east-bank fishway or by constructing a new fish bypass channel around the east side of the dam. These recommendations are in agreement with some of the key recommendations made by the

U.S. Fish and Wildlife Service in their three action study programs aimed at implementing solutions to fishery problems at RBDD.

Even with these recommended actions, it is doubtful that manipulation of RBDD operations, within the constraints of present and proposed future water demands, will ever completely reverse present losses. Strictly from a fishery standpoint, the logical solution to RBDD fish passage problems would be to replace the dam with a pumping plant to supply water to the Tehama-Colusa and Corning canals. At the Glenn-Colusa Irrigation District, a pumping plant similar in size to the one that would be required at Red Bluff was installed at a cost of $10,000,000 in 1984. If RBDD is not to be replaced with a pumping plant, or another source of water is not developed that would allow raising the gates, a formal agreement should be made relative to raising the gates at least during the nonirrigation season to improve fish passage.

Until studies demonstrate that ripe salmon handled at the RBDD trapping facility spawn successfully in the river if released, they should be spawned artificially and their spawn placed in the USFWS incubation station constructed for that purpose. Operation of this facility should be funded by the Bureau of Reclamation, owners and operators of RBDD. Moreover, studies should be intensified to develop a positive plan for eliminating squawfish predation at RBDD. Finally, the city of Redding's proposed Lake Red Bluff Power Project should be opposed unless all the fish protective measures recommended by DFG and USFWS are incorporated in the project.

Chapter Nine

The Sacramento River Winter Chinook Salmon

Threatened with Extinction

Jack E. Williams and Cindy Deacon Williams

The Sacramento River winter chinook salmon are nearing extinction. The unacceptable loss of this distinct and valuable race of salmon would be the result of conscious management decisions that demonstrated a lack of concern for the needs of the species. The winter chinook salmon are adapted to entering the main Sacramento River in late winter and spawning far upstream during the early months of the Central Valley's long, hot summer. Their ancestral spawning grounds were in the McCloud River, a tumbling, spring-fed tributary of the upper Sacramento. Eggs hatched and fry matured in the cold, consistent flows of the McCloud, seemingly oblivious to the hot summer weather.

All this changed when Shasta and Keswick dams were built on the Sacramento. Migrating adults, blocked by the dams, no longer could reach historic spawning areas. Pollution, water diversions, and stream channelization also exacted their toll. As recently as 1969, more than 100,000 spawners were tallied. Annual counts from 1982 to 1988 average only 2,334 adult fish—more than a 97 percent decline. At the reduced population levels of recent years, extinction is likely from continued habitat losses or a chance event such as drought or flood.

This disastrous decline has called fishery biologists, anglers, and

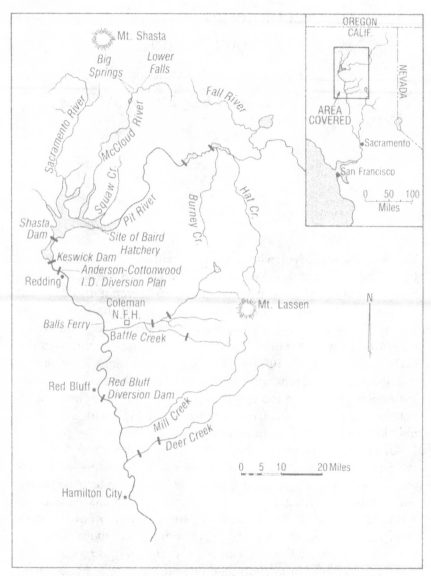

Map 2 Upper Sacramento River and Tributaries

environmentalists to action. They began efforts in 1985 to protect the winter chinook salmon pursuant to federal and state endangered species acts. Faced with the precedent of possibly listing a salmon as endangered or threatened and having to restrict water development in the Sacramento Valley, the National Marine Fisheries Service and the California Fish and Game Commission hesitated to apply the protections afforded by strong endangered species laws and ultimately did so only after extensive legal wrangling and a further precipitous decline to only five hundred and seven spawners in 1989. This chapter traces the unique life history of the Sacramento River winter chinook, efforts to save them from extinction, and the anticipated impact of endangered species protection on sport and commercial anglers, water users, and the fish.

Biological Background

Four races of chinook salmon occur in the Sacramento River system: fall, late fall, winter, and spring runs. The runs are distinguished by timing of adult upstream migration, spawning, egg incubation, and juvenile downstream migration. Most adult winter chinook typically move upstream in December through March and spawn in May and June. Eggs incubate during summer months when water temperatures often are critically high. Downstream migration of winter chinook juveniles occurs from early August through October. In addition to temporal separation of the runs, the winter chinook are further distinguished by their choice of spawning gravels in depths of about ten feet and by their younger age at the time of upstream migration. Historically, winter chinook were mostly three-year-olds, with a few four-year-olds, whereas other runs had a larger proportion of the older fish. Other distinguishing features are their relatively low fecundity, rapid upstream migration of adults, and extended staging period of adults in headwaters before spawning. For these reasons, the winter chinook are considered to be racially distinct from all other runs of chinook salmon.

State of the Resource

Spawning winter chinook salmon were first observed in the McCloud River in 1902 at the site of the Baird Hatchery. In 1942,

To be extinct? This winter-run Sacramento River chinook salmon, having traveled more than two hundred miles upriver without feeding, still appears ocean-fresh. The hardy species, which originally populated the McCloud River drainage, is nearing extinction. (California Department of Fish and Game)

however, Sacramento River salmon were blocked from reaching McCloud River spawning habitat by the construction of Shasta and Keswick dams. Neither dam has fish passage facilities. At first, winter chinooks were nearly eliminated because successful spawning was impossible in the warm Sacramento River water below the dams. By 1946, however, releases of cold water from the depths of Shasta Reservoir were cool enough to allow successful egg incubation. By the late 1960s, cool-water releases from Shasta and Keswick dams maintained winter chinook spawning

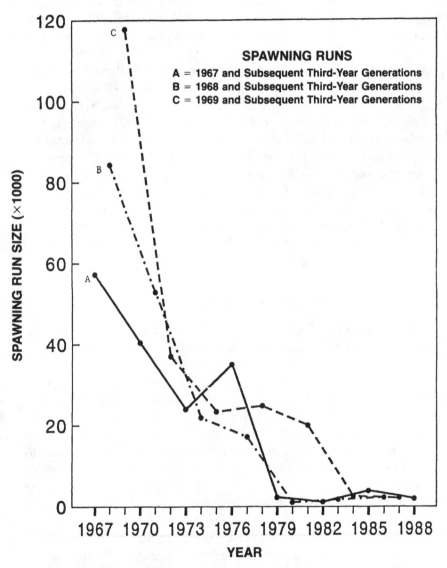

Size of 1967, 1968, and 1969 winter chinook spawning runs passing Red Bluff Diversion Dam. Subsequent generations are plotted on three-year cycles to reflect typical ages of spawning adults. (California Department of Fish and Game, unpublished data)

runs of fifty thousand to one hundred twenty thousand fish. Runs have declined dramatically since that time (see graph).

Many other problems plague the winter chinook: pollution from agricultural runoff, toxic waste from Iron Mountain Mine, gravel mining in tributary streams, channelization and bank stabilization of the Sacramento River, and obstructions to migrating adults and young created by the Anderson-Cottonwood and Glenn-Colusa irrigation districts' diversion dams. Decline of runs was further exacerbated by construction of Red Bluff Diversion Dam in 1966. Water below the dam is rarely cool enough for successful spawning, and fish must ascend fish ladders in order to reach cooler waters released by Shasta and Keswick dams. During the first nine years of operation of that facility, each three-year generation of winter-run fish ascending the ladders declined by one-half. Random environmental events, such as the 1976–1977 drought and the 1982–1983 El Niño, further depressed the winter chinook populations. The 1976 spawning run of 35,096 adults, for example, produced a spawning run of only 2,364 fish three years later.

A Threatened or Endangered Species?

With spawning runs of the winter chinook salmon reduced to only a few thousand fish, the California–Nevada chapter of the American Fisheries Society petitioned the National Marine Fisheries Service in November 1985 to list the run as a threatened species pursuant to the Endangered Species Act of 1973, which establishes that "species" can include "any subspecies of fish or wildlife or plants, and any distinct population segment of any species [of] vertebrate fish and wildlife which interbreeds when mature." The Sacramento River winter chinook salmon thus could be listed as a threatened or endangered species. Furthermore, recent studies elsewhere on the West Coast have documented the genetic distinctiveness of various salmon runs among and within large river systems.

After a review of the winter chinook's status, the National Marine Fisheries Service determined in February 1987 that listing was "not warranted." Although the service found that the run qualified as a species as defined by law and that numbers of winter chinook had seriously declined, it determined that listing was not warranted because resource agencies had informally agreed to a ten-point conservation plan. Major provisions of the plan include:

1. The U.S. Bureau of Reclamation would raise the gates at Red Bluff Diversion Dam to facilitate passage of winter chinook adults.
2. Gravel would be introduced to supplement riverine spawning habitat.
3. A winter chinook propagation program would be initiated at Coleman National Fish Hatchery.
4. A temperature control curtain would be constructed at Shasta Dam.
5. Iron Mountain Mine would be cleaned up with Superfund money.
6. Fishery restoration plans for the Sacramento River would be developed pursuant to California Senate Bill 1086. (This bill established an advisory committee to develop and recommend recovery actions for Sacramento River fish, wildlife, endangered species, and riparian losses.)

In February 1988, the Sierra Club Legal Defense Fund on behalf of the American Fisheries Society and others filed suit in U.S. District Court to force listing of the Sacramento River winter chinook as a threatened species. The plaintiffs argued that because the run is in fact biologically threatened, the federal government, under the Endangered Species Act, has a nondiscretionary obligation to list the fish. They argued further that the incomplete and nonbinding nature of the planned conservation measures was insufficient to prevent extinction of the species, especially if a drought were to occur or if additional habitat were lost to hydroelectric projects or other human endeavors.

Meanwhile, the Sacramento River Preservation Trust petitioned the California Fish and Game Commission in September 1986 to protect the Sacramento River winter chinook pursuant to California's Endangered Species Act. In June 1987, the commission rejected the petition, refusing to consider the need to protect the winter chinook. After a coalition of eight environmental and sport-fishing groups filed suit in state court, the attorney general advised the Fish and Game Commission to accept the petition for consideration. The commission did so in February 1988, granting "candidate" status under the state's Endangered Species Act for one year to provide time for further study.

Late in 1988, pursuant to a stipulated agreement filed with the court, the National Marine Fisheries Service reconsidered their "not warranted" determination, mainly because of poor water conditions resulting from low runoff during the previous winter. The service formalized aspects of its ten-point plan by signing an agreement with the U.S. Fish and Wildlife Service, the Bureau of Reclamation, and the California Department of Fish and Game that mandated voluntary compliance but included no penalties for failure to perform. Thus the agreement failed to guarantee that its conservation policies would be implemented fully. The National Marine Fisheries Service nonetheless reaffirmed its negative decision. Concluding that the agreement was seriously flawed, the American Fisheries Society urged the court to set an early trial date on the merits of the lawsuit.

On November 7, 1988, all parties met in U.S. District Court in Sacramento to make their arguments to Judge Raul A. Ramirez. Opening discussions focused on the request of the Pacific Legal Foundation, representing a number of water users and districts, to intervene on the side of the National Marine Fisheries Service. Ramirez, characterizing the request as one based solely on economic interest and therefore untenable under the Endangered Species Act, denied the motion to intervene. The main ruling, however, went against the salmon. Judge Ramirez ruled that the National Marine Fisheries Service has the authority to determine a species' status and declared he would not interfere. The American Fisheries Society immediately appealed the trial court decision, and in April 1989 the salmon case was joined by a case involving the listing of the spotted owl that had been heard in U.S. District Court in Seattle before Judge Thomas S. Zilly on November 9, 1988. Although the spotted owl case involved a different species and a different agency (the Fish and Wildlife Service), the legal specifics were the same. In both situations an attempt had been made to avoid listing by developing some form of conservation plan. Nonetheless, the respective judges had made opposite rulings.

Meanwhile, disturbing news was coming in from the field. Engineering analysis by the Bureau of Reclamation was calling into question the feasibility of the critical temperature control curtain proposed for Shasta Dam. With estimated costs jumping from $5 million to $50 million, the project was sent back to the proverbial drawing board. Early estimates of the 1989 run size also were worrisome, indicating approximately five hundred and sixty winter chinook. The final number ultimately dropped to five hundred and seven adults. Further, the Fish and Wildlife Service was able to capture only forty-two of the hundred adults desired for artificial propagation at Coleman National Fish Hatchery.

By early May 1989, the outlines of a disaster at Coleman were becoming clear. More than half of the winter-run adults captured had died before spawning because of antibiotic reaction, fungal infection, and furunculosis. Ultimately eggs could be taken from only one female. The final three survivors were returned to the river in the hope that they might spawn under natural conditions. Serious questions arose about the hatchery's ability to mitigate recent losses.

With the dramatic decline in population size from the two thousand adults that many had argued was the new "stable" population size, the California Fish and Game Commission reversed its February 1989 decision not to list under California law and on May 16 declared that the winter run should be listed under state law as an endangered species. Pressure was building at the federal level, too. In a surprise announcement during a briefing conference on May 25, attorneys for the National Marine Fisheries Service conveyed the service's intent to list the salmon. In the August 4, 1989, issue of the *Federal Register,* an emergency rule was published that granted threatened status to the winter chinook under federal law. Given the earlier decision by the Fish and Wildlife Service to list the spotted owl, both potentially landmark cases regarding an agency's obligation to list became moot. The legal answer awaits another day.

Effects on Sport and Commercial Anglers

If the Sacramento River winter chinook salmon were to be listed as *endangered* pursuant to federal law, ordinary taking of the fish (killing, capturing, harming, or angling for it) would be a violation of that law. But since it is to be listed as *threatened,* as the American Fisheries Society has requested, take agreements can be developed or special regulations added to allow for sport and commercial harvest in accordance with other applicable regulations. Numerous precedents exist for regulatory flexibility in such cases.

Even more harvest flexibility is present in California's Endangered Species Act. The Fish and Game Commission may authorize sportfishing of any species listed as endangered, threatened, or a candidate. Also, the state act exempts commercial take if it complies with other state laws and regulations.

The National Marine Fisheries Services has noted that winter chinook probably are subjected to a lower ocean harvest rate than other runs of Sacramento River salmon. Because their adult migration is temporally separate from the more numerous fall-run chinook salmon and their average size is smaller, few winter chinooks are harvested commercially. Sportfishing has less impact, as well, but because large inriver take could reduce population numbers unacceptably, the Fish and Game Commission has established a "rolling closure" on this fishery as adults migrate upstream.

Recovery of the Winter Chinook

Chances for survival and eventual recovery of the winter chinook should improve once listing pursuant to federal and state endangered species acts is finalized. Perhaps the most significant consequences of listing under either act would be requirements for interagency consultation. The federal act requires all federal agencies that authorize, fund, or carry out projects affecting the threatened or endangered species to consult with federal fishery offices on possible adverse effects. Projects likely to be subject to this requirement would include bank stabilization work by the U.S. Army Corps of Engineers, operation of all Bureau of Reclamation dams on the Sacramento River, and new hydroelectric projects licensed by the Federal Energy Regulatory Commission. State agency consultation with the Department of Fish and Game will similarly be required when their projects affect the resource.

Projects seldom are stopped through consultation requirements, but they occasionally are modified to mitigate harm to the threatened or endangered species. A 1987 General Accounting Office study found that no western water project was terminated as a result of consultation. Operations of Shasta Dam and Red Bluff

Diversion Dam are key factors in providing suitable temperatures for spawning, egg incubation and rearing, and allowing fish access to water of favorable temperature. The Bureau of Reclamation, without pressure from the Endangered Species Act, has yet to guarantee adequate conditions for the winter chinook. In 1988, for example, the bureau raised the gates at Red Bluff Diversion Dam to improve fish passage but prematurely lowered them when conflicts with water contracts became apparent.

Propagation of the winter chinook at the Coleman National Fish Hatchery is a tempting way to recover the run. As has been shown, however, implementing a successful winter chinook propagation program will not be easy. Already more than $2 million has been spent to build deep-water holding ponds at Coleman National Fish Hatchery to accommodate the prolonged period that the winter chinook hold in spawning areas prior to egg maturation.

Numerous other technical problems are likely to be encountered as the effort continues. Some problems are more complex. Salmon produced at hatcheries often are biologically or genetically inferior to wild fish. Researchers from the University of California at Davis have documented that hatchery chinook salmon were more vulnerable to predation by Sacramento squawfish as they pass Red Bluff Diversion Dam than were wild chinooks. Severely depleted anadromous salmon stocks may lack the genetic plasticity necessary to adapt to culture or later readapt to natural environments. Impacts on the success of natural reproduction also must be evaluated before individuals are removed from the wild population to establish a hatchery stock.

Restoration of the Sacramento River ecosystem would be of enormous benefit to other salmon runs in the river. The spring chinook, for example, is seriously depleted from historic levels and fast approaching the need for protection under the Endangered Species Act. In the preamble of the act, Congress declared that "various species of fish, wildlife, and plants in the United States have been rendered extinct as a consequence of economic growth and development untempered by adequate concern and conservation." The act has the tools to rebuild the winter chinook salmon by restoring habitat and reducing mortality factors. Listing is only the first step. It's time we put every available tool to its intended use.

Chapter Ten

What's a Salmon Worth?
Philip A. Meyer

Value is a broad concept, capturing diverse perceptions and concerns. Indian elders link the survival of salmon with survival of their tribe as a people. Commercial fishermen value the fact that salmon enable them to live in a place and manner special to them. Sportfishermen often use the term "priceless" to describe a salmon striking in the solitude of a breaking dawn or the warm experience of fishing with family or friend.

Such values extend beyond fishing people. Residents of the San Francisco Bay area and Sacramento, most of whom do not fish, assign hundreds of millions of dollars of value to maintaining viable salmon stocks in the Sacramento/San Joaquin system and passing them on in good health to future generations.

All these perceptions are valid, but such material, cultural, and philosophical perceptions of the value of salmon and steelhead are often balanced against other alternative uses of California's streams and coastline. This balancing is often done in terms of dollars, and it typically results in decisions adverse to these salmonids. Here I want to discuss such "economic" measures of value—what is and isn't measured and how manipulations by economic analysts often make economic values appear small or large.

What Can Economics Measure?

At its narrowest, economics deals with the exchange of goods, services, and resources in the marketplace. It describes costs of pro-

This remarkable underwater video shot shows a chinook salmon in the Pacific Ocean off the Golden Gate just before striking a sportfisherman's bait. (Dick Pool)

duction, revenue received from sales, and resulting income and employment for persons who produce and sell. Thus the economist's product should be useful for commercial businesses associated with salmonids—either those catching, processing, and/or selling fish or those providing services to sportfishermen.

But problems emerge in this market value process because most of the human activity associated with salmon and steelhead in California is not dictated by markets. Heir to a legacy of open access to natural resources in the American West, sportfishermen incur only nominal access costs, and the values that residents associate with knowing California's salmon and steelhead are "alive and well" are not directly marketed at all. This has led economists to develop a second level of evaluation: "nonmarket" valuation. Nonmarket valuation attempts to estimate the dollar worth of sportfishing for salmon and steelhead as if, in fact, such fishing opportunities were bought and sold in markets. Economists employing this device test their assumptions about what motivates human behavior by actually observing sportfishing activities and questioning sportfishermen or residents directly.

Nonmarket valuation techniques are only one yardstick of value. The ultimate value is how salmon and steelhead affect people's material, social, and psychological well-being in California. The dollar yardstick can never capture more than some of this value. Several fundamental reasons for such undervaluation are discussed below.

Market economic values inevitably understate total values for salmon and steelhead in California. Even when nonmarket valuation techniques are incorporated in economic analysis, it is doubtful that dollar economic estimates can ever deal fully with the value of salmon and steelhead in the state. Dollars simply cannot capture every cultural, aesthetic, or spiritual value that Californians associate with salmon and steelhead. Consequently, the key issue for those who consider economic value is not whether a combination of market and nonmarket values is all you need to know, but rather what component of value can realistically be described in dollar terms.

Economists have traditionally used four procedures which ensure that economic values for salmonids will be a small component of total value. These procedures do not reflect Californians' values so much as they do the untested assumptions built into economic models. The first two procedures are based on a simplistic belief that Californians are a pretty homogeneous people lacking significant diversity in motivations, beliefs, preferences, and circumstances.

Thus, on the business side, some economists assume that all Californians are very mobile; if a person can't get a job on the North Coast, for example, he need only pack up and get a job elsewhere. In fact, if the person does not move, he is seen, in this view, as less worthy. Although such inferences obviously do not fit the facts, economists holding such a view reduce business values associated with salmon and steelhead by up to nine times—or ignore them entirely.

On the recreational side, some economists expand the economic assumption that Californians are a pretty homogeneous lot to assert that ready substitutes for salmon and steelhead sportfishing exist. This assumption recently led one economist to consider a loss of over nine thousand California salmon as not very serious and to underestimate the associated nonmarket impact on sportfishing by up to twenty times. Despite the lack of empirical evidence, some economists adhere to this position, and salmon and steelhead continue to be undervalued.

A further assumption of some economic calculus is that "beyond ten or twelve years, the future doesn't matter." Thus some economists discount future benefits at seven percent or more, reducing salmonid values to trivial amounts after a decade. This practice affects small streams severely. Yet, if properly cared for, many such streams are capable of producing moderate levels of salmon and steelhead in perpetuity. Again, such practice is based primarily on economists' assumptions, not observable reality. Californians do care about future generations. This practice, however, while still widespread, is diminishing.

Finally, some economists assert that significant losses of salmonids in California will have value effects equivalent to those for significant gains—and inquire how much California's commercial and sportfishermen "would pay to avoid losses" caused by water export and other instream degradation. This posture defies overwhelming empirical evidence that economic estimates of losses to salmon and steelhead that are based on fishermen's "willingness to pay" for mitigation underestimate salmon and steelhead value by as much as twenty times. Psychologists tell us that people will value losses of salmon and steelhead fishing they now have more highly than additional fishing successes they might obtain. Further, adherents of the "willingness to pay" economic school seem to defy recent public trust doctrine decisions confirming Californians' right to fish.

In sum, economic evaluation of California salmonids should not be expected to produce a complete accounting of the values Californians associate with these fishes. Dollar value estimates nonetheless play an important role in determining salmon and steelhead futures—particularly when other activities deleterious to salmon and steelhead are being considered. Also, the assumptions implicit in an economist's analysis often have a determining effect on both the comprehensiveness and the magnitude of salmon and steelhead values reported. It is consequently incumbent upon those contemplating use of economic values for salmon and steelhead that they ask several questions of the economist:

How does he view the question of population mobility? Does his analysis assume that fishers readily move from place to place and consequently unemployment in certain areas is a short-term phenomenon—or, because fishermen prefer their particular lifestyle, is a degree of unemployment seen as a chronic condition?

Does economic analysis assume existence of ready substitutes for sportfishing—or does it recognize a degree of uniqueness of California's sportfishing opportunities?

Does the analysis give proper weight to long-term salmon and steelhead values—or does it seem to suggest that future values do not matter?

Does economic analysis assume that California's salmonids are or should be for sale to the highest bidder—or does the analysis assume that Californians who value salmonids should be adequately compensated for losses?

Only when these questions have been asked and answered can citizens and decision makers determine whether a particular economic analysis adequately represents the values they hold relative to California's salmon and steelhead.

Present Estimates of Economic Value

The general role of economics in representing value for California's salmon and steelhead has been shown. Key issues that will determine the relevance of an economic study have been identified. One further qualification must be borne in mind: even the best estimates of the importance of salmon and steelhead to Californians are far from complete. Most economic estimates to date have considered value at a point in time and have not adequately addressed the issue of the long-term importance of salmon and steelhead to California's fishermen, businessmen, and Indian peoples who depend upon them for material survival—nor their importance to the small communities where these groups often live.

The estimates presented here in tabular form represent a "snapshot" selected from recent work by the author but do not address these important broad issues. These estimates assume that:

1. Commercial fishermen, fish processors, and businessmen supporting sportfishing cannot easily change their occupation or location. Revenue generated by salmonids for these sectors is consequently significantly beneficial.

2. Sportfishing locations and opportunities are scarcer in California than they were a decade ago and have a degree of uniqueness. Conversely, substitute locations and activities are not freely available.

3. Californians seek a better balance between salmon and steelhead benefits for themselves and future generations beyond the next ten to twelve years—and use discount rates reflecting long-term concerns.

4. Adequate compensation is the proper measure for valuing losses to salmonids—or in valuing restoration of previous losses. The proposition that losses to California's salmonids should be valued according to what fishermen would pay to avoid them is incorrect.

Contemporary estimates of the economic value of California's salmon and steelhead, expressed in dollars per fish, are presented in Table 1. Value for the full present population of California salmon and steelhead is presented in Table 2. Again, these values are incomplete and do not incorporate full sustaining occupational and life-style benefits for commercial businessmen, other dependent California residents, or Indian peoples. Values associated with the "existence" of salmonid stocks and passing them in good shape to future generations (bequest values) are not included. Existence and bequest values for Sacramento/San Joaquin River chinook salmon alone were recently calculated to be several billion dollars annually. It should be noted that commercial and sportfishing values have been evaluated using differing economic procedures of differing comprehensiveness. Estimates of sportfishing value are more comprehensive. For this and other reasons, such estimates cannot be compared, fish by fish, with commercial values.

In sum, California's salmon and steelhead values are substantial—with the greatest portion of that value lying outside commercial markets. As economic methodology improves and is more realistically applied, the component of total value that can be captured by

TABLE 1   *Economic Value (in dollars) per California*
Salmon or Steelhead

	Watershed			
Value	Sacramento/ San Joaquin	Klamath/ Trinity	Eel	Other Coastal Rivers
Chinook Salmon				
Commercial fishing	23.59	20.15	22.44	22.44
Commercial fish processing	12.26	10.47	11.66	11.66
Commercial fish retail	11.12	9.50	10.58	10.58
Total commercial fishery value	46.97	40.12	44.68	44.68
Net revenues to businesses serving sportfishermen	21.84	31.20	31.20	31.20
Sportfishing nonmarket value	675.00	179.00	179.00	179.00
Coho Salmon				
Commercial fishing	0	9.80	9.80	9.80
Commercial fish processing	0	5.04	5.04	5.04
Commercial fish retail	0	4.62	4.62	4.62
Total commercial fishery value	0	19.46	19.46	19.46
Net revenues to businesses serving sportfishermen	0	31.20	31.20	31.20
Sportfishing nonmarket value	0	179.00	179.00	179.00
Steelhead				
Net revenue to businesses serving sportfishermen	39.94	24.96	24.96	24.96
Sportfishing nonmarket value	530.00	530.00	530.00	530.00

Source: Meyer Resources, Inc., April 1988, "Benefits from Present and Future Salmon and Steelhead Production in California."

economic estimation will increase. This progression is important if prior economic underestimation of the value of salmonids is to be reduced. Even under ideal conditions, however, the decision maker must recall that economics provides only one measure of a broad range of values associated with salmon and steelhead in California.

TABLE 2 *Economic Value (in millions of dollars) of Salmon and Steelhead Stocks in California*

River System	Net Benefits Per Year		Total Present and Future Net Benefits	
	Business Benefits	Business Plus Nonmarket Benefits	Business Benefits	Business Plus Nonmarket Benefits
Sacramento/ San Joaquin	19. 7	101. 4	886	4,561
Klamath/ Trinity	6. 8	23. 5	306	1,057
Eel	3. 0	16. 1	135	727
Navarro	0. 1	1. 0	6	47
Carmel	0. 05	1. 1	2	50
Ventura	0. 025	0. 6	1	25
Other Calif. rivers	0. 35	3. 4	16	156
All Calif. rivers	30. 0	147. 1	1,352	6,623

Source: Meyer Resources, Inc., April 1988, "Benefits from Present and Future Salmon and Steelhead Production in California."

Chapter Eleven

The Human Side of Fishery Science
Dave Vogel

The desire to understand natural phenomena and an intense fascination with fish and water are fundamental reasons one becomes a fishery field biologist. To better understand fish, it helps to "think like a fish." Such a thought passed through my mind as I narrowly cleared an enormous boulder while being towed upstream along the bottom of the Sacramento River, thirty feet below the surface. Clinging tightly to our underwater planing boards and stretched out like Superman in flight, my buddy diver and I waved about in the fierce, erratic current like the long strands of green moss growing on the riverbed. As the turbulent water in the deep, narrow river channel pummeled us, we hoped our scuba gear was well fastened.

Our U.S. Fish and Wildlife Service field crew was conducting a study below Keswick Dam to map out the salmon spawning gravels and the distribution of deep-spawning winter chinook salmon in the three-and-a-half-mile river reach between Redding and Keswick. The purpose was to determine the potential impact a proposed mainstem hydroelectric power plant might have on upstream salmon spawning habitat.

When the U.S. Bureau of Reclamation's Shasta and Keswick dams were completed in the mid-1940s, chinook salmon and steelhead lost half of their historic spawning grounds in the Sacramento River basin and were forced to spawn only in areas downstream from the dams. These downstream spawning grounds are relentlessly disappearing, however, because the dams sealed off the river's principal source of gravel. Results of this underwater survey would ultimately reveal just how much valuable spawning gravel remained following more than forty years without significant replenishment.

A deep, throaty hum of the powerful V-8 engine in the large jet boat somewhere above us was barely audible as it pulled us slowly along the riverbed darkened by the considerable depth. A quick glance upward revealed a disconcerting lack of the familiar shimmer of sunlight experienced in shallower depths. We could see only the tow ropes and radio communication cables bowed into the swift current and disappearing up into the pea-green, low-contrast water.

Sights no other human had ever seen unfolded: enormous and unusual bedrock outcroppings sculpted into convoluted, ethereal forms over aeons by the raging torrent from the immense upper Sacramento River watershed. These sights had been seen only by fish such as chinook salmon and steelhead as they skillfully negotiated the river's hydraulics on their upstream migrations, enticed by the smell of their natal waters where they would complete their life cycle.

Conducting field investigations of this nature isn't new to the U.S. Fish and Wildlife Service. Fifty years before this spawning gravel study, Fish and Wildlife Service field investigations were being conducted at almost the same location. Those earlier scientists were studying salmon and steelhead run sizes to develop baseline information for evaluation of the potential impact of Shasta and Keswick dams before and during their construction.

The Reasons for Field Studies

There are numerous national policy statements and congressional authorities by which the U.S. Fish and Wildlife Service conducts field investigations. The following are just a few examples.

The Fish and Wildlife Act of 1956 authorizes the development and distribution of fish and wildlife information to the public, Congress, and the president, as well as the development of policies and procedures that are necessary and desirable to carry out the laws relating to fish and wildlife. This includes steps required for the

development, advancement, management, conservation, and protection of fishery resources. Results of fishery studies can become
an integral part of this information.

The Fish and Wildlife Coordination Act, first enacted by Congress in 1934, is one of the oldest environmental review statutes. Its
primary purpose is to ensure that fish and wildlife resources receive
"equal consideration" among aspects of resource development. This
act requires that any water resource developer (federal agency or
other party subject to federal regulation) must consult with appropriate federal and state fish and wildlife resource agencies. In all cases,
this includes the U.S. Fish and Wildlife Service; the service presents
the project sponsor with recommendations on how to avoid negative
impacts on fish and wildlife resources. Results of field investigations
are often used to develop these recommendations. If adverse effects
cannot be avoided, recommendations are made for proposed compensation or mitigation of damages.

The Fish and Wildlife Service functions similarly for reviews of
hydroelectric projects licensed by the Federal Energy Regulatory
Commission (FERC) under the authority of the Federal Power Act.
Upon request by FERC, the service conducts environmental review
before project construction begins. In these instances, needed information on baseline resource conditions is often developed through
field investigations. After review of the data, the service sometimes
recommends that the water project not be constructed because its
impact would be too detrimental to fish and wildlife resources and
could not be sufficiently mitigated. Examples include two proposed
hydroelectric projects on the mainstem Sacramento River near Redding and Red Bluff. After extensive review, both the Fish and Wildlife Service and the California Department of Fish and Game found
that these two hydro projects would cause major damage to the
salmon and steelhead resource that would not be mitigated by actions proposed by the license applicant. Sometimes, however, the
responsible federal regulatory agency decides otherwise and the
project gets built.

Salmon and steelhead stocks of the Central Valley and the
Klamath and Trinity basins are among the highest-priority, nationally significant fishery resources in need of restoration. Because of
the depleted status of the runs in these watersheds and because of
major federal water development impacts, the service has a signifi-

cant responsibility to assist in the restoration of these stocks. In March 1988, the service released a promising plan to implement the agency's fishery program responsibilities:

1. To facilitate restoration of depleted, nationally significant fishery resources
2. To seek and provide for mitigation of fishery resource impairment due to federal water-related development
3. To assist with management of fishery resources on federal (primarily service) and Indian lands
4. To maintain a federal leadership role in scientifically based management of national fishery resources

Field studies will play a key role in implementing these responsibilities.

Technological Advances

In conducting field investigations, modern-day fishery biologists benefit from a recent proliferation of sophisticated mechanical and electronic devices to assist them in data collection. Elaborate fish trapping devices allow scientists to capture fish alive and release them unharmed. Recent advances in scuba technology now permit biologists to study salmon and steelhead directly without significant danger. The advent of video and major improvements in underwater communication devices have greatly increased the ease and accuracy of conducting fishery investigations in the watery realm of salmon and steelhead. When direct underwater observations are not feasible, technological advances in miniaturization and electronics enable scientists to attach small tags or radio transmitters to fish and study their life history characteristics.

These technological advances in field equipment can sometimes lead to unexpected findings in biological studies. Service biologists experienced an example of this when they were using directional antennas to monitor downstream migration behavior of radio-tagged six-inch-long juvenile steelhead. Over the course of the study, a curious behavior pattern emerged among the young steelhead. Some of the fish would stop at one point in their downstream migration near an island and reside there for several days

Modern technology at work. This U.S. Fish and Wildlife Service biologist is monitoring movements of radio-tagged salmon below the Red Bluff Diversion Dam. (Dave Vogel)

until the small battery inside the transmitter was drained and radio signals ceased. After the first one or two occurrences of this unusual behavior, it was assumed that the radio transmitter had fallen off the fish and was on the bottom of the river against the bank, beeping away while its porter continued its seaward journey.

But as this pattern continued, biologists became suspicious and decided to make a closer examination of this mysterious site where the transmitters apparently chose to abandon their free ride to the high seas. Precise record keeping by the biologists showed that, in all cases, the transmitters were pinpointed to a very specific location near the uppermost tip of the island, where a cluster of large, dead cottonwoods hung over the water. Down among the root wads in the river, transmitters were accumulating at an alarming rate. When biologists saw that the dead trees were home for an extensive cormorant rookery, the enigma was solved. The hapless young steelhead migrating past the cormorant nests carried on their backs shiny transmitters that caught the sharp eyes of the always-hungry piscivorous birds. Radio tags were subsequently camouflaged to reduce that hazard.

Fish Tagging

Overall, fish tagging is one of the most powerful tools a biologist has to study salmon and steelhead. Tagging enables biologists to study many biological facets of salmon and steelhead. Information on their early life history, migratory characteristics, distribution and movements, harvest rates, age and growth, mortality rates, and population sizes can all be obtained from fish tagging data.

The development of miniaturized coded-wire tags opened many doors of fishery research that were previously closed. These tags, about the size of the broken-off tip of a sewing needle, enable scientists to study fish from their fry and juvenile life stages up to their adult phase. Every year, hundreds of thousands of juvenile salmon released from state and federal hatcheries in California carry these small tags. When the young fish is only two or three inches long, a microscopic metal tag with a unique binary code is harmlessly implanted inside the snout cartilage. At the same time, the small adipose fin on the back of the fish is excised. This fin, about the size of a large coin on an adult salmonid, is not regenerative. Therefore, when a fish is later captured (for example, by an angler fishing in the ocean) the absence of an adipose fin indicates that a coded-wire tag is probably present in the fish's snout. After the fish is caught by an angler or returns to a hatchery, a biologist can then remove the tag from the fish by using a metal detector, a microscope, and surgical equipment. The binary code on the tag establishes the origin of the fish, the exact time and location of its release, and its size when released. This tool has enabled fishery biologists to dramatically improve the survival of fish released from hatcheries by determining the best time, size, and location to release them.

Despite many success stories, biologists have learned that tagging studies don't always bear fruit. During an initial attempt to artificially propagate Sacramento River winter-run chinook at Coleman National Fish Hatchery, nearly twenty thousand young salmon were tagged with coded-wire tags by Fish and Game and the Fish and Wildlife Service and released to obtain biological information on the least understood run of chinook in California. Not a single one of the fish was ever seen again. What fate befell them may never be known. Understanding this dangerously depleted stock of salmon remains one of the greatest challenges to salmon biologists today.

But sometimes biologists glean unanticipated information from fish tagging studies. During a cooperative study between the Fish and Wildlife Service and Fish and Game, reward tags were attached to steelhead smolts released from Coleman National Fish Hatchery to determine how many of the catchable-sized trout were caught by anglers during the steelhead's migration to the ocean. One of the reward tags was returned from an angler fishing near the American River two hundred miles downstream from where the fish was released. Surprisingly, the angler wasn't fishing for trout. He'd found the tag attached to an eight-inch-long young steelhead inside the swollen stomach of a hefty striped bass.

A Fish Eye's View

Despite all the technological advances in field investigations, there can be no substitute for simple, direct observations. This was well exemplified during the service's study of the impacts of the U.S. Bureau of Reclamation's Red Bluff Diversion Dam on salmon and steelhead. The primary purpose of the dam, which was put into operation in 1966, is to provide irrigation water for the Tehama-Colusa and Corning canals. A fish bypass system was constructed to route young downstream-migrating salmon and steelhead away from the canals and back into the Sacramento River. The point where these fish were jettisoned into the river resembled five small geysers boiling and spewing water just downstream from the dam.

Bureau of Reclamation engineers called the exit structure in the river a fish bypass terminal box. From the surface, only a spectacular turbulence could be seen because the structure was deep underwater. A quick peek down to the bottom of the river in the vicinity of the bypass terminal box seemed essential to biologists in order to understand its effects on fish. It wasn't difficult to understand why anyone hadn't done so earlier because the site was intimidating— even terrifying—for the most adventurous scuba diver.

Once submerged, our diving team encountered disorienting, pounding turbulence and blinding air bubbles coming from all directions. We quickly found refuge from the angry boil in a back eddy on the downsteam concrete wall of the terminal box. We held our position there momentarily, feeling like meteorologists making observations from the eye of a hurricane. Tightly grasping jagged boulders for leverage, we slowly crept along the river bottom and cautiously poked our facemasks around the corner of the wall's safety and peered into the maelstrom. A shocking sight leapt into view.

Large, decaying salmon carcasses, their heads caught in heavy vertical steel grates, fluttered in the powerful current exiting the structure like pieces of ribbon tied to a cooling fan. The cause of the problem was immediately apparent: adult salmon migrating up the river were attracted by the high-velocity water surging from the structure. They attempted to swim into the terminal box and became fatally trapped when they rammed their gills into the four-inch-spaced vertical steel grates.

The engineers' "terminal box" now took on a whole new meaning. The structure had been there for more than twenty years and had been killing adult salmon every year without anyone having the slightest inkling of the deathtrap on the bottom of the river. When asked about the purpose of the four-inch grates, engineers replied that it is the optimal spacing for trash and debris deflection at dams and other man-made structures in the river. When informed of the severity of the fishery problem the bureau agreed to cut out alternate grates, thereby making eight-inch openings to allow unimpeded movement of fish in and out of the terminal box. Subsequent underwater fish tagging operations showed that adult salmon could then come and go as they pleased without physical injury. A crude solution, but effective.

Engineering versus Biology

After several years of scuba diving in the Sacramento River around the fish bypass terminal box, Fish and Wildlife Service divers began to believe that one of the five bypass pipes had less flow compared with the other four. Because of the extreme turbulence and the divers' inability to get close to the pipes, direct measurement of water velocities couldn't be performed, but it was clear that something wedged inside that one pipe must be restricting the flow.

Knowing that millions of young downstream-migrating salmon and steelhead must pass through these pipes every year, biologists confronted bureau engineers with their concern. After hours of poring over the original bypass blueprints and engaging in heated discussions, the engineers were adamant: it was impossible for anything to be wedged inside the fish bypass system because the inside of the pipes was "smooth as silk and contained no sharp bends." "After all," they maintained, "who would know better than the engineers who had designed the system?"

Ultimately, the service's field biologists prevailed with their arguments and persuaded the bureau to shut the pipe down, seal off upstream and downstream openings, crack open a manifold they said had never been opened, and conduct an inside inspection of the drained pipe. Feeling a little guilty over the apparent hassles we'd created for them by insisting on an action they believed would be fruitless, I volunteered to be the "spelunker" and crawl up through the several-hundred-foot-long, thirty-inch-diameter pipe. My less than eager offer was eagerly accepted.

Access to the manifold was buried deep underground in a musty, dark concrete chamber, much like a 1960s cold war bomb shelter. Once the manifold was cracked open, the subterranean chamber was filled with a putrid, eye-watering, gut-wrenching smell. "That's just because it was sealed off for so long," bureau workers said as I donned a diving suit for a long belly crawl. With the echoing roar of motors high overhead pumping fresh air down into the manifold, I grabbed a flashlight and extra batteries and squeezed inside the narrow pipe. The anticipated long journey lasted no more than several feet up into the pipe.

Just a short distance from the manifold opening, wedged tightly inside the pipe, was a grotesque accumulation of tree limbs, assorted river debris, and fish and animal (cats? otters?) carcasses. The entire mass was jammed up against three large steel vanes welded to the inside of the pipe.

According to an old-timer with the bureau who'd been there since the fish bypass system was constructed (sheepish because he'd forgotten it until now), the steel plates were added shortly after construction as "flow straightening vanes" to improve the accuracy of water velocity measurements. Engineers' flow meters had long since been removed, but the steel vanes were left. By the end of the week, the bureau had cut all the vanes from the five pipes and ground the surfaces smooth.

Working Conditions in the Field

During those cold winter evenings when driving rain beats against the windows and bone-chilling wind howls through the trees, most sane people confine themselves to the safety and warmth of their homes. Under these conditions, many a fishery biologist is shivering in the field, out on the rivers and streams—home to California's salmon and steelhead during the freshwater phase of their life cycle. It is then that nature allows scientists to unravel some of her mysteries.

Field investigations conducted after sunset have revealed some of the more interesting characteristics of young salmon and steelhead growing and migrating in California's streams and rivers. Nighttime field investigations have shown that nature tells the fish to migrate downstream to the ocean at night to avoid such hazards as sight-feeding predatory fish and birds. This simple fact prompted biologists to recommend that the bright lights routinely found on dams and diversion structures be turned off at night to reduce nocturnal predation on small salmonids.

Biologists have also learned that many of the young salmon and steelhead move downstream during times of winter turbidity. Efforts to acquire this particular item of information on salmon and steelhead biology involved a near disaster. The winter of 1983 was unusually wet, and many streams and rivers in northern California approached or exceeded record flows. Service biologists had recently constructed a new river trawling apparatus designed to capture young fish live during their downstream migrations. The logistical problems in using the device on a flooding river seemed immense but not insurmountable.

Fishery biologists, more often than not, have to work on a shoestring budget and make do with very limited resources. The result is equipment, such as this novel sampling gear, built by improvisation, ingenuity, and scrounging for parts. Hardly recognizable as a boat, it was a floating complex of metal frames, cables, winches, pulleys, planing boards, and a large pile of fine-mesh net. Several rigorous tests during low river flows had provided convincing evidence that it would perform well under any condition. Thus when the rivers were rising to flood stage that winter, we took to the field—confident that essential new data would be acquired.

With anticipation in the air as on Christmas Day when a child tries out a new toy, the trawl net was spooled out midriver into the roiling, muddy floodwater. The cold, driving rain added excitement to the moment. The trawl boards opened the net in the current, steel cables twanged taut—and the trawl boat suddenly lurched backward under the strain. Something was terribly wrong. Forward momentum ceased and the vessel with its three wide-eyed biologists, mouths agape, was swept uncontrollably downstream. Scrambling, fumbling fingers flipped toggle switches and activated winches to retrieve the trawl. The electric motors immediately growled out their disapproval under the tremendous load.

The trawl net was eventually retrieved after a harrowing ride with the boat full speed in reverse in order to chase the swollen net downstream to ease the strain on cables and winches. Close examination of the net revealed that extremely small pieces of vegetation invisible in the muddy floodwater had plugged the small mesh the instant the net opened, creating a sea anchor hell-bent on sweeping the boat out to the ocean. Debris accumulation was expected, but over a course of minutes, not seconds. In those brief moments, though, much to our howling delight, dozens of inch-and-a-half-long salmon had accumulated in the live trap at the end of the net. We learned several lessons from that experience.

The Reward

Salmon and steelhead field biology is a tremendously challenging profession in this environmental age. The greatest challenge is the urgent need to solve seemingly unsolvable problems.

With proliferation of water development in California, competing demands on salmon and steelhead freshwater habitat have never been more intense. Procedural decisions guiding development require sound biological data on affected fish and wildlife resources, and good data require sound field investigations.

The ultimate reward is seeing an increase in runs of these magnificent fishes and knowing that findings from field investigations contributed in some measure to their resurgence.

Chapter Twelve

Women and Fishing on the North Coast

Mary-Jo DelVecchio Good

"I'm not a fisherman's wife, I am a fisherman."

—Lee O'Bryant

Although women are seldom associated with California's commercial salmon fishery, their energetic participation in numerous voluntary and political organizations, such as the Salmon Restoration Association (SRA), the Noyo Women for Fisheries (NWFF), and the Pacific Coast Fishermen's Wives Coalition (PCFWC), has enhanced the vitality of the fishery and the industry. Yet it is as "fishermen," particularly as salmon trollers, that women have recently crossed the historical barriers of gender in the world of commercial salmon fishing. In this essay, the experiences of women who fish or who are partners in family fishing enterprises at sea and on shore give insight into the range of roles, not always evident, that women provide fishing communities.

On Being Fishing Women

Although it is no longer unusual to find women in fishing partnerships with men, when Betty Roberts first began to fish with her

A version of this chapter appeared originally in *Ridge Review* (Summer 1983). Special appreciation is expressed to Jayne and Rich Bush from the Noyo salmon fleet and to the women who so willingly gave their time for these interviews.

Commercial salmon fishing, once solely a male occupation, now includes women, such as this one fishing off Cordell Bank. Fishermen's wives typically are part of a fishing team and often go to sea with their husbands. (Marie De Santis)

husband in 1963, women were a rarity in the commercial salmon fleet. Betty recalled the cultural barriers that existed at that time to women fishermen, even those who fished with their husbands:

The old-time fishermen believed it was bad luck for a woman to be on a boat. . . . They really believed it. I don't know why they name their boats after women! It really upset them that I went fishing with Vernon. They wouldn't fish near us, talk to us, and weren't very nice, period! Vernon said to them, "My wife's going to go fishing whether you like it or not. This is a free country. I don't believe your stupid superstitions." And so I just kept going with Vernon and pretty soon we were catching more fish than they were. By the end of the season all of them were talking to us. They enjoyed me being out there . . . and hearing a woman's voice on the radio.

Cultural barriers, such as traditional beliefs that women on a boat bring bad luck to a fleet, were overcome in one season for Betty; however, it was and continues to be much more difficult for women to be regarded as "fishermen" in their own right. Women who fish either independently or as equal partners with their husbands still find it necessary to distinguish that they are "fishermen" and not fishermen's wives.

Lee O'Bryant, who had fished for over a decade with her husband, Wayne, spoke graphically of being a "fisherman," not just an assistant to her husband. As the first female board member of the Noyo Salmon Trollers Marketing Association (1983), she earned public recognition for being a "fisherman" from her male salmon troller peers. Yet in our interviews, Lee felt the need to assert that *her* experiences of fishing, the sense of independence and competitiveness, the sense of being "dedicated or crazy," were also common to other salmon trollers—to *male* fishermen. She related: "I think all fishermen have to be a tiny bit touched just to be fishermen! And that comes from one. It's one of the last frontiers . . . being able to be your own boss, go when you want to go . . . only they are stopping that too."

Lee also captured the competitive facet of fishing, usually associated with men but clearly experienced by women as well, especially by women who make decisions of when to go out and where to fish: "When it comes to where you're going to fish now—grrr . . . I'm a real competitive person, and if someone else's got fifty fish, I want to know why I don't have fifty fish. I don't want to be in the middle of the bunch. I want to be in the top! I'm not. Very seldom."

Dobie Dolphin, a woman in her early thirties who had recently entered the fishing fleet at Albion, had quite a different experience from that of the older women "fishermen" who worked out of the more traditional Noyo harbor. Dobie became involved in fishing, first as a diver, placing and finding moorings, and second as a boat hand, pulling fish. It was the excitement and satisfaction of pulling fish that led her to think, "Maybe *I* could have a boat." She received "good support" from the Albion fishing community in her endeavor. Albion has a noted counterculture community of migrants to the coast, and "tradition" is neither powerful nor restrictive.

Thus women have crossed the boundaries that traditionally separated the ocean world of fishing and fishermen from the home communities and fishermen's wives.

Dividing the Tasks

Women and men who fish in partnership often divide up tasks in such a way that the men do most of the pulling, of bringing the fish in, and the women run the boat and cook. Although both partners may ice and clean the fish, it is this common division of labor that allows for the persistence of the distinction between "fishermen" and the women who are their partners. Although some women become pullers (with a "decline in the complexity of meals prepared when working in the pit," as Printha Platt noted), and fish equally with their partners, others first experienced the exhilaration of catching fish while their husbands were asleep. Betty Roberts recounted how she came to be a "salmon puller" one afternoon while her husband was asleep and the fish weren't biting:

I was going around in circles and doing all these goofy things . . . and all of a sudden these lines started going. Boy, they were really pulling on the springs. So I put the pilot on . . . I was trying to be really quiet—so Vernon wouldn't wake up. I was going to pull the fish and surprise him. I tell you, I got the line up and this huge fish was on there. And I could see there were two more fish on the line. Big fish. And I thought, oh my goodness, what do I do now! Well, I got hold of the leader and I started bringing it up. And I went to gaff it and I threw the gaff away. It just went flying out of my hand. So I netted that fish! Back there, by myself, which is hard to do. And I *wrassled* that thing onto the boat. And I got this other one coming up and I got it into the net and I threw it on the boat. And I look up, and there comes Vernon, asking, "What the hell are you doing?" I said, "I'm catching fish!" He couldn't believe it. Here's these two really pretty ones and a whole bunch more on the line. And he said, "Well, let me get them," and I said "NO! They're mine!" I think we had about eight fish from that rigging at that time, and then they really got going. That's what really did it. I really had a ball that day!

Women on the ocean also contribute in unique ways to the fishing enterprise. The women from the fleet who had fished in the past or were currently fishing felt that women frequently play an integrative role in the commercial fisheries. Betty Roberts speculated that a woman's presence on a boat changes the relationships among fishermen, makes them more cooperative, more united, less cliquish, more politically aware. This integrative role at the personal level carries to the political and community level, through activities in the political and voluntary organizations, such as the SRA, the NWFF, and the PCFWC:

Just a women being there, they couldn't be man-to-man . . . it's kind of like our coalition. We have women from all the states (Washington, Oregon, California), and we have helped all the fishermen to get along. Because we brought things out in the open. We discussed things. We helped each other out with all of our problems—where men would not do this. We would go home and talk to our husbands and inform-them about things. A fisherman was a fisherman, and all he thought about was going out there and catching fish. But they have a lot more to think about today.

The Fishing Relationship Ashore

On land or sea there is a bittersweet aphorism current in the fishing community that a successful fisherman is one who has a working wife. Many wives have become full partners in the fishing endeavor. Others are involved in fishing through the support they give their husbands on land, by running errands, buying groceries, picking up gear, assisting with boat maintenance, and managing finances. "Turnaround time," the time spent in ports between trips, is reduced by such activities.

It was as a full-time fishing couple that Printha Platt and her husband missed the support activities other fishermen's land-based wives provided—"We missed that wife, missed her very much."

Women who are involved in their own careers, who provide a steady income against the seasonality and vagaries of the fishing

income, find that being married to fishermen can at times be trying. Printha, who gave up fishing for a career ashore and engage-
ment in fishing politics, commented on the new season:

I feel far less anxious approaching this season as a fisherman's wife from
shore than I have as a fisherman. . . . Under the right circumstances, yes,
I'd rather be fishing. To me, the right circumstances are—you know the
old joke about the fisherman. If someone gives him a million dollars, he
uses it until it's all gone!

Management of the fishing relationship from ashore requires
that women also be independent, and many are resilient and move
between the two worlds of being involved in fishing and committed
to careers of their own. Printha conveyed this sense:

I know when the guys are gone, I need my separate life, and make myself
as flexible as possible. . . . Before I went fishing with Buzz, whenever he
came in from a trip . . . , I'd drop everything. I ran my life that way. I'd
call my boss and say, "I'm not going to be in for two days, the boats are in!"
You give up a lot of your own life. The paradox of it all is that in order to
survive as a fisherman's wife, as a wife who is left behind, you have to have
something to make your own life worthwhile.

The unpredictability of their husband's occupation poses difficulties for women who are maintaining the family household and caring for children. Lyndsey Miller expressed that uncertainty:

The hardest thing for me is never knowing from day to day, from minute
to minute, if Larry is going to be fishing or not. There are a lot of things
that I am doing now that I like doing, but the children have to be taken
care of. . . . I feel strongly that Larry should be with them . . . and of
course he's not during fishing season. So that's really the hardest thing.
Never being able to plan anything. But on the other hand, it would be
very tiresome to always know what you are going to do. Fishing is a
strange one!

Being on the Ocean

Although "fishing is a strange one," being on the ocean is compelling for many women, be they "fishermen" or "fishermen's wives."
Jayne Bush, who sustains the partnership with her fisherman husband through her involvement in the Noyo Women for Fisheries
and the Salmon Restoration Association and her professional career

on shore, still relishes the fishing days when the beauty of the ocean overwhelmed and "the sunsets consumed the sky."

Fishing women talk about the ocean with fondness, as if it were the most natural place to be. The language of challenge, often used by men speaking of the sea, rarely arises spontaneously in women's conversations about the ocean. Women more often speak of the ocean in idioms of endearment, often with a depth of passion, as well as in terms of beauty. Dobie Dolphin explained her "love" for the sea this way: "I am an ocean person . . . a double Scorpio. I would much rather be out on a boat than on land. . . . It is the movement, everything is in flux, exciting, special."

Susan Bondoux, who had owned her own thirty-two-foot Monterey, compared her love of the ocean as "nearly stronger than the love I have for my children." She reflected on the call of the ocean, even in fishing seasons wrought with economic and legal difficulties:

From the first time I went out there, I was never all right after that. . . . Everything is magic out there. . . . The ocean has always been a real thing for me. Peace. It's always peaceful to me. Being out there seems the most peaceful, most beautiful thing I've ever done. And things can happen really fast. . . . I'm not afraid, I have never been afraid. Whether it's lack of good sense, I have no idea. . . . It's rockin' in your Momma's arms. You anchor up at night. And everything is so beautiful. I feel safe out there.

Concluding Note

Fishing is largely a family business, and wives of fishermen have traditionally made important contributions by maintaining stable home lives, providing shore support for fishing activities, and more recently by devoting time and energy to political activities and restoration efforts required to make the industry viable. It is by going to sea as a "fisherman," however, that women have crossed the traditional boundaries between male and female, boats and home, and life on the ocean and life in the community. As a result, traditional roles have been reorganized on some boats, women have entered business and political activities as owners, and a particular sense of relation to the ocean has enriched the cultural language of fishing. With a special passion for the sea, for fishing, and for the industry, these women contribute to the vitality of the salmon trolling enterprise.

Chapter Thirteen

The Lower Klamath Fishery

Recent Times

Ronnie Pierce

During the 1800s, progressive white settlement of California had drastically disrupted Indians' lives. One event at the close of that century had a tremendous effect on the Klamath/Trinity fishery: in 1894 a through road from Eureka to Crescent City was completed, and it crossed the Klamath estuary. With improved access, the Klamath River soon gained a reputation as one of the finest sport-fishing rivers on the West Coast. In addition to commercial fishing, Indians could now supplement their income or earn seasonal living as guides for sportfishermen.

A succession of poor salmon runs in the late 1920s led sportfisher-men, through the Klamath River Anglers Association, to initiate state investigation of causes. Even though non-Indian commercial salmon trollers out of Eureka were fishing off the mouth of the Klamath, investigative committees appointed by the legislature in 1929 and 1932 determined that the decline was caused by Indian commercial river fishermen. They recommended closure of that fishery.

In 1933, after several years of controversy, the state of California

Space limitations dictated this chapter's focus on the currently controversial Yurok gillnet fishery of the lower Klamath River. It should not be inferred, however, that other Indian tribes of the area have not been involved as well. The Hupa tribe, particularly, has exerted tremendous effort over the years to define and protect Indian fishing rights.

Last of its kind. This cannery at the mouth of the Klamath River, owned and operated by non-Indians, employed Indian fishermen and plant workers. It was closed by law in 1933. (Peter Palmquist collection)

closed the commercial Klamath River fishery, which had been operating since 1876, and reinstated the Klamath as a navigable waterway. Opponents of the ban on commercial fishing felt that the closure would deprive many Indians of their livelihood. Proponents countered that the Indians could make more money as guides and boat pullers for tourists. *The National Waltonian,* August 1933, reported: "Sportsmen the country over will rejoice that the Klamath, famed for its piscatorial delights . . . has been saved for the public."

The Yurok fishermen took no such delight as the state of California, unchallenged by the federal government, banned both commercial sales and the use of gillnets by Indians for subsistence fishing in the lower Klamath. Indians lost more than their jobs— their most efficient method of fishing for food was made illegal. The propriety of state control of Indian fishing made manifest in this 1933 action was not legally addressed for several decades.

In 1969, the California Department of Fish and Game seized the gillnets of Raymond Mattz, a Yurok fisherman, and sought superior court permission to destroy or sell the "illegal" nets. Mattz claimed that state code prohibitions against net fishing did not apply on an Indian reservation. Although the state won in the lower court, Mattz appealed. The California Supreme Court upheld the lower court's judgment, but the U.S. Supreme Court, in *Mattz v. Arnett* (1973), reversed that judgment. In this unanimous decision, the Supreme Court held that the lower twenty miles of the Klamath River was still a reservation, despite its having been settled by non-Indians.

The case was then remanded to the trial court for a ruling on the applicability of state regulation to Indian fishing. After several appeals by the state, the U.S. Supreme Court, in *Arnett v. Five Gill Nets* (1976), upheld the rights of Indians to fish on the reservation free from state regulation.

The cases evolving from the confiscation of Mattz's nets created a jurisdictional void in the matter of Indian fishing regulations on the reservation. The Bureau of Indian Affairs (BIA), as the federal trustee of Indian resources, moved to fill that void. In a key decision, the U.S. solicitor general concluded that "we know of [no authority] that would limit an Indian's on-reservation hunting or fishing to subsistence. The purpose of the reservation is not to restrict Indians to a subsistence economy, but to encourage them to use the assets at their disposal for their betterment. . . . Moreover, Indian fishing rights have on several occasions been interpreted . . . as extending to commercial fishing."

In 1977 the BIA published its first regulations to govern Indian fishing on the reservation. These allowed restricted commercial fishing in addition to subsistence fishing, but they were vague as to the amount of commercial harvest allowed and determination of which Indians were legally qualified to fish.

Improved regulations were published for the 1978 season, but the clamor resulting from the reinstatement of the Indian gillnet fishery continued to grow. A new sportfishing organization called the Klamath River Basin Task Force, as well as Del Norte County and the Hoopa Valley Business Council, initiated action to place a moratorium on commercial Indian gillnet fishing pending completion of an environmental impact statement.

A "strike force" of federal agents and park police, numbering up to seventy-five men, in a hostile confrontation closed the Indian net fishery in midseason 1978. This "conservation moratorium," as it came to be known, lasted until 1987. Subsistence fishing was still permitted, and some violations of BIA regulations were reported during the period.

During the years between the Indian commercial fishery closure in 1933 and the mid-1970s, conservation had indeed become a key word in relation to the Klamath. Conservation measures were obviously needed. Intense logging, construction of major dams, and burgeoning offshore and inriver fishing took their toll. Two natural disasters, the 1964 flood and the 1977 drought, added to the destruction. The once fabled salmon runs of the Klamath River had become a national concern.

In 1976, the Pacific Fisheries Management Council (PFMC) was created by the federal Fishery Conservation and Management Act. Its primary role is to develop and monitor biological management plans for fish stocks from three miles to two hundred miles off the coast of Washington, Oregon, and California. This territory includes the Klamath Management Zone (KMZ), the commercial fishing area where Klamath River salmon are most vulnerable to harvest.

The initial PFMC management plan for the river called for a spawning escapement (the number of fish permitted to reach spawning grounds) of 115,000 fall chinook. From 1979 through 1982 the average actual number was 35,000 fish. Rather than close the ocean fishery, which would devastate local economies, the PFMC decided to institute a long-term correction, a phased twenty-year rebuilding schedule. The plan did not work: after only two years, spawning escapement had declined to 22,700 fish.

With the further decline of stocks because of the 1983 El Niño, PFMC felt its only option for 1985 under the rebuilding schedule was to prohibit ocean commercial harvest of salmon in the Klamath Management Zone. The impact of this action led to the formation of a "Klamath River Salmon Management Group" under the auspices of the PFMC. Representatives of lead agencies, with representatives of all affected fishing groups, met at the table—many tables!— to negotiate. By March 1986, they had worked out an agreement.

This innovative, nonadjudicated agreement included a new management scheme for the river. The new plan, called harvest rate management, was based not on a fixed spawning escapement goal but on percentages of maturing adult fall chinooks that could be harvested by all fishermen—ocean or river—with a fixed percentage of the year's mature stock being left to spawn. Conservation responsibilities would be shared equally among user groups. Because its mathematical formulas promised to consider the needs of all groups, including the fish, it was guardedly acceptable to everyone involved. The resulting formal agreement was to be in effect for five years.

Under the new plan, the ocean fishermen were allowed a limited fishery in the KMZ in 1986, and in 1987 the Indian fishermen finally achieved an allocation of fish that was sufficient to lift the "conservation moratorium" placed on them in 1978. They were legally allowed to harvest fish commercially for the first time in fifty-four years (discounting the ill-fated effort during the late 1970s).

In its second year, however, the new plan had to deal with problems not fully anticipated during negotiations: migrating salmon stocks from many separate river systems mix in the ocean and therefore contribute at different rates to the ocean harvest. Current technology dictates that the contribution rate of Klamath-origin salmon can only be estimated after the season's closing. That rate, a natural biological function, fluctuates according to location of catch effort and relative strength of stocks from other contributing rivers.

Before the season opens, biologists and regulatory agencies can only estimate what the abundance and contribution rate of the Klamath salmon will be, and from that estimate they shape a season to determine a harvest goal for Klamath fish. If their ocean population or contribution rate predictions are in error, serious harm to the fishing economy or the spawning escapement can result.

If the Klamath contribution rate is underestimated, ocean fishermen in the predetermined season may inadvertently take many more Klamath salmon than the allocation agreement permits. In such a situation, as occurred in 1987, with total abundance and contribution rates underestimated, both ocean and river fisheries are allocated fewer fish than accurate data would have allowed. The problem for Indian fishermen in such a case is that although the salmon catch of the river fishery can be accurately tallied—every fish entering the Klamath River is a "Klamath fish"—Indian harvest is still restricted to the erroneous preseason predictions. While ocean trollers may harvest fish in other zones when their KMZ allocation is met, Indian river fishermen have no such compensating option.

Thus problems remain. While ocean regulations severely limit fishing within the KMZ, areas north and south of the zone have relatively liberal seasons. Fishermen in these areas of mixed-stock fishing take almost all of the Klamath salmon allowed for ocean harvest under allocation agreements, leaving few to allocate to their fellow fishermen inside the zone.

No one is happy. Ocean fishermen are distressed with limited seasons, and agency biologists are frustrated with the demand, both biological and social, for management numbers they are unable to supply. Nor are the Indian fishermen happy. Most perceive any increase in the allowable ocean harvest, while they in the river fishery must conform with the numbers of the agreement, as yet another breach of trust.

Negotiations continue on overall harvest rates and allocation. Ocean trollers want more Klamath fish to harvest in the KMZ. Their counterparts to the north and south are unwilling to reduce harvest of other stocks to protect Klamath fish mixed with them. The Indians, with their revitalized commercial fishery economy, will certainly fight to keep their allocation. And, above all, a spawning escapement must be developed and maintained that will allow for perpetual renewal of stocks.

While conservation efforts, through negotiation and regulation, are working to a certain extent, as evidenced by increases in spawner escapement, companion efforts in the restoration of fishery habitat in the Klamath/Trinity basin must also be addressed. Indian people residing on the river from its headwaters to the ocean have long decried the destruction of the spawning and rearing habitat of the river. Impacts of the modern world are many: logging that denudes hillsides and dumps silt onto spawning beds; dams that reduce the river's flow; streambanks stripped of trees, which causes overheated water; unscreened diversions for agricultural irrigation that bleed off juvenile fish; road-crossing culverts that block passage to spawning beds; mouths of streams filled in with deposited gravel, blocking fish passage. All these conditions contribute heavily to the decline of fish production.

Young fishermen, ageless site. Karok tribesmen hoopnetting for salmon at the Klamath River's Ishi Pishi falls in 1989. This fishing site has been used by Native Americans since ancient times. (Andy Kier)

Life cycle nearly complete. These salmon are migrating to spawning areas in the Trinity River. Most salmon return to their native waters to spawn. (Bureau of Reclamation)

Cool water needed. Dredging equipment removing gravel to create cold water pools for protection of salmon and steelhead on the Trinity River. (Bureau of Reclamation)

In 1982, the Bureau of Indian Affairs responded to concerns about declining salmon runs by commissioning the Klamath River Basin Fisheries Resource Plan. The plan provided the needed insight and direction for congressional action to restore Klamath River fishery resources. Legislation enacted in October 1986 provides that $21 million will be appropriated by the Department of Interior from October 1, 1986, through September 30, 2006. A matching $21 million is to be provided by nonfederal sources. Under the act, a fourteen-member Klamath River Basin Fisheries Task Force has been created that embodies all pertinent management agencies, fishing groups, and counties of the basin.

Tribal management of tribal resources is the goal of Indian people nationwide, and the U.S. government, through the Bureau of

Indian Affairs, promotes this goal of Indian self-determination. The Yurok people, as they now move toward the twenty-first century, have managed to hang on to their fishing rights and have reestablished the right to sell a portion of their harvest.

On October 31, 1988, Congress created a separate Yurok Reservation and actuated formal organization of the Yurok tribe. The right to restore and manage the fishery resources of the reservation, both "in the gravel" and at the policy level, has always been a high priority of Indian fishermen. This most recent action has the potential for future resolution of the confusing jurisdictional issues that have been inherited. The stage is set for great strides in Indian self-determination, as well as restoration of the Klamath basin's fishery resources.

Chapter Fourteen

The Commercial Troller
Bill Matson

Since ancient times, those hardy men and women who "go down to the sea in ships" have held a mystique for us all. The sleek lines of a ship capture the imagination as one studies her, and few can escape the temptation to imagine themselves at the helm, guiding her out to the open sea.

On a typical Sunday afternoon all across America, waterways are filled with boats of all kinds and sizes, as modern man tries to capture a bit of the sea's adventure and beauty—and perhaps escape for a while into that simpler world where nature makes all the rules and man's tasks are reduced to humble compliance and skilled navigation.

There is something alive about a ship on the open seas as it rises and falls with the swell. After riding out a storm, man and ship seem somehow bonded together in a rugged union of appreciation for each other. Each new storm and each passing year strengthen that bond until seamen and ship find contentment only in the gentle roll of the open ocean.

The commercial troll fishermen from the beginning has been a blend of seaman and fisherman. Although diesel engines have replaced the sails, the call to the sea remains unchanged. The earliest trollers used sail power and pulled their heavy trolling lines by hand to land a fish. Even after gasoline engines became common after the turn of the century, hand lines were used for more than twenty years until the modern powered gurdy was invented. To-

Commercial salmon troller preparing to land hooked salmon. Note power winches (gurdies) that raise and lower troll lines. (Marie De Santis)

day, efficient diesels have replaced the old undependable "one-lunger" gasoline engines. Electronic equipment to allow navigation in fog and at night makes the ocean safer, and improvements in troll gear, especially monofilament leader, make fishing more effective. But the ocean remains the same.

The demands upon men to navigate, operate cranky equipment in all kinds of sea conditions, and still manage to catch fish have produced seamen as fine as any in our great past. Commercial fishermen are rugged individualists who daily must contend with unexpected weather changes, deciding where to go when the fish disappear, maintaining and repairing equipment, and cooking their own meals while the galley floor keeps shifting.

Accountable only to themselves, dependent upon their own skills at reading ocean conditions and catching fish, it is little wonder these men find the complexities of bureaucratic procedure unnecessary nonsense unrelated to the real issues.

It was mid-July 1969 and life seemed uncomplicated. Everything was good: good fishing, good weather, good time to be alive.

The ocean was flat this morning, I observed with satisfaction, as we left the Humboldt Bar just after first light. No swell at all today.

I set my compass heading for NW and relaxed as my little troller, *M&M*, steadily put the bar behind. I thought I would go out to about sixty fathoms to start with and then go from there. Reports from other fishermen were encouraging, so I looked forward to a good day. Not many kings had been reported, but silvers were abundant, so as I ran I looked for silver signs—surface feeding. I wasn't disappointed.

Not long after passing the fifty fathom line, I began to see fish on the water. I could see other trollers working outside of me, scattered over a wide area. Before long, I slowed down and set my gear. Fishing was pretty good at first, and I quickly caught thirty to thirty-five silvers. Then, as the sun climbed higher, the fish quit biting but remained on the surface, often jumping right out of the water.

For the next three days the pattern was the same: I would catch a few fish in the morning and late evening, and endure long, unproductive hours between. All through those long days, I watched fascinated as I passed through acres and acres of salmon. They were finning near the surface, they were jumping out of the water—but they would not take a hook.

As I worked my way north from Eureka to Crescent City, the pattern was the same: a hundred miles of silver salmon on the surface of the water!

But it wasn't all idyllic. On the second day out I sighted my first Russian trawler. He was working among the salmon fleet, dragging his nets just south of Redding Rock. Later I found out there were several such factory ships in the area. That same day the skipper of the *City of Eureka*, a large trawler out of Eureka, tried to catch up with the Russian vessel, but soon learned it was impossible with his boat. Even towing nets, the Russian ship moved faster than *City of Eureka* could go at full speed with no nets out.

That was my introduction to the speed, size, and efficiency of the foreign midwater fleet. Convinced that these foreign vessels were fishing for salmon, irate fishermen approached Congressman Don

Clausen for protection. The cry for a two-hundred-mile economic zone became our theme as the best solution for keeping the foreign fleet out and providing opportunity for the American fishermen. When the two-hundred-mile bill (the Fisheries Conservation and Management Act, 1976) passed, we were confident this meant long-term security for the American coastal fishing fleets.

Most of us in the fishing industry know that catching large numbers of fish is not hard at all if that is the only concern. Large, powerful boats with nets are capable of quickly cleaning out the fish resources of the Pacific Ocean's narrow continental shelf. Salmon are especially vulnerable to overfishing because of the ease with which nets can harvest them and their delicate dependence upon healthy streams in which to spawn.

The troll industry has no interest in depleting the resources by overfishing. We are committed to fighting for restoration of salmon. We also believe trolling to be the best method of harvesting salmon. Here's why.

First, by its very nature, trolling is inefficient. The chances of overfishing with troll gear are slim because of the finicky fish that do not always bite a hook, the small size of the boats that can economically afford to troll, and the unpredictable weather conditions on the ocean.

Second, ocean harvest always includes at least two year classes: three- and four-year-old fish. This reduces the impact on the older, spawning class and thins the population to some extent, which functions to improve conditions for survival of remaining fish—fewer fish competing for the same food supply. Even if not caught by hook and line, the overwhelming majority of salmon that enter the ocean as smolts do not survive to return to their home river.

Third, the fish are harvested during the ocean phase of their lives, providing the highest-quality product possible.

Fourth, harvest opportunity is spread over the widest area, providing economic value to dozens of communities along the coast and thousands of fishermen.

In recent years the troller has felt irate at times over management philosophies that emphasize harvesting of upmigrating adults at river mouths (called terminal harvest) and management that fails

A bustling commercial herring fishery survives in San Francisco Bay. Many commercial salmon fishermen harvest herring, crab, and other saltwater species when salmon season is closed. (Marie De Santis)

to act upon real issues facing the salmon in its efforts to survive. Management of the resource so far has really been management of the fishermen. Only superficial efforts have been made to increase streamflows, reduce water temperatures, solve gravel recruitment problems, and improve other elements of habitat. A good example is offered in the report of the Klamath River Salmon Management Group Allocation Committee In-River Caucus Platform, December 15, 1986: Goal Number 1 is "to shift towards more of an *inside* harvest of the Klamath River stock between Ocean and In-River users." Of six stated goals, only one addresses enhancement of runs, and even then in only a general way. Four of the six address a single issue: how to divide up the pie for harvest.

If managing the fishery resource comes down to a fight over how large each user group's share should be, none of us has a future: not

Spud Point marina in Bodega Bay was constructed in 1985 solely to accommodate commercial fishing vessels. (Earl Carpenter)

Short season. In 1988, offshore commercial salmon trollers in the Klamath Management Zone caught their quota within four days of opening. Most Eureka-based fishermen had to move south to complete the season—or sell their boats, such as the F/V *Pelican*. (Mark Lufkin)

Aerial view of Lewiston Dam and Fish Hatchery on the Trinity River, built to mitigate loss of salmon and steelhead habitat associated with the construction of Trinity Dam. (Bureau of Reclamation)

the troller, not the Indian, and not the sportsman. And certainly not the salmon.

A management scheme that starts by focusing on habitat, from spawning gravels all the way downriver to the ocean, and then manages harvesters by regulating gear, number of fishermen, and seasons, will provide fish for everyone. It will also provide healthy streams, forests, and wildlife areas for future generations.

A management scheme such as that of the past eight or nine years, which pits users against each other, insists upon using bad data to predict fish abundance, and cannot respond to economic need or obvious calculation errors—such a system is doomed to fail miserably. And as that failure develops, a whole industry will be lost, a priceless resource depleted, non-Indian and Indian alienated from one another.

Chapter Fifteen

Rivers Do Not "Waste"
to the Sea!

Joel W. Hedgpeth and Nancy Reichard

Flowing rivers are part of the circulation system of our globe. They bring nourishment to the land and to the estuaries in which they end as they pass into the sea, where their last waters evaporate from the surface of the sea and become again the falling rain and snow that renews them. Their waters as they flow are the life of the growing plants along their banks and of the fishes and other beings that live in them. They carry with them the sediments that enrich the land and help the waters carve their channels and their banks. Their ultimate effect is to wear down the mountains and level the earth.

The hydrologist Luna Leopold has estimated that there are about three million miles of river channels in the United States, but beyond a few of the greatest rivers of the globe it is difficult to rank streams because their three principal dimensions—length, area of drainage basin, and volume of annual dishcarge—are independent of each other. But no matter how rivers are ranked, the streams of the Central Valley of California fall far below on a scale of magnitude, and even in California the combined drainage area of the Sacramento and San Joaquin rivers is little more than three-fourths the area of the Klamath River drainage (9,676 to 12,000 square miles), although their combined average annual discharge is almost fifty percent greater, that is, 21.8 million acre-feet to the Klamath's 12.9 million acre-feet. California's largest rivers are two orders of magnitude smaller than the Columbia. Probably no river system in the country is under the stress of greater use in proportion to its size than California's Central Valley rivers.

Rivers are not just sources of water to be used by the most destructive organisms yet to inhabit the earth; in their natural state they are dynamic equilibrium systems that have characteristic morphologies in common. For example, the bed load of sediments is roughly in balance: the sediment load carried away during floods is rebuilt during low flow periods so that on the average there is a tendency for scour to be balanced by fill. A river, apparently, can only stay straight so long, not much more than ten channel widths, and there is a constant relationship between the width of a river channel and its meander radius. In his studies, Leopold emphasized that the intermediate flows of the river, because they are the most frequent throughout the year, typically have the greatest cumulative effect on the landscape they move in.

Massive dams on a river upset this natural equilibrium of the movement of sediments as they accelerate deposition in the still waters above the dam and assist the unburdened waters below to cut back the natural banks, especially where vegetation has been removed from the banks and floodplains, as on the Sacramento River. Where the drainage is in forested terrain, logging activities may add to the sediment load and shorten the life of the reservoir downstream.

Our civilizations were first built along the rivers, the Nile, the Tigris and Euphrates, and the world's great cities are river towns, Rome, Paris, London, New York. Indeed, a city without a river is incomplete. In our times we have learned to bring the rivers to the cities by great systems of dams, tunnels, canals, and pipes from distant mountains and irrigate vast fields to raise food to support them. Too often the engineers and business people who have made these developments possible have forgotten the real nature of rivers, the effects their changes and withdrawals may have on the structure of the rivers, and consider that any water which escapes their purposes is "wasted."

The fish that live in the rivers are the last thing water developers think about. This is especially true in California, where the great dams being built before World War II were started without any concern for the fish. How was it that Shasta Dam was designed and partly constructed before anyone remembered the salmon? It was planned only for power and irrigation; even when "amenities" were thought of, fish were remembered last. People and government agencies focusing on irrigation wanted to have nothing to do with any other purpose.

A Case in Point

Although the dynamics of river systems may vary from region to region, a look at the potential effects of the Dos Rios high dam project on the Eel River, a proposal much in the news in the 1960s and 1970s, illustrates many of the problems associated with river diversions. (That proposal is currently dormant because of protections by wild-river acts, but developers still look longingly at the Eel River basin, with its thirty-seven hundred square miles of drainage area, as a potential source of exportable water. A change in political will could strip North Coast rivers of their current protections.)

California's northern coast is a geologically unstable area. Its rivers, such as the Eel, are unique in that rates of sediment production from their watersheds are greater than those of any other region of comparable size in the country. Because of this, impacts due to diversion of water from this region may be significantly different from those associated with similar projects elsewhere. A major diversion on the Trinity River substantially hurt fishery resources of that stream, and complex and costly attempts to mitigate the impacts have not yet proved successful. The Eel, with a headwaters diversion into the Russian River, has likewise been impaired, but on a smaller scale.

The most serious effect of a new dam on the Eel River would be a reduction in sediment transport capacity of the river downstream of the diversion. Large volumes of sediment are dumped into the river from tributaries and streamside landslides. High flows needed to flush this sediment downstream would be totally eliminated by a large project. Accumulation of sediment in the river would cause spawning gravels and pools used by salmon and steelhead to become choked with fine sediment, as has occurred in the Trinity River. As the streambed rose, bank erosion would become more frequent, and the resultant threat to streamside property would tend to offset any flood protection benefits. The whole river would be changed.

The Eel River ranks second statewide in production of steelhead and coho salmon and third in production of chinook salmon. A large water diversion project would make inaccessible one hundred fifty or more miles of anadromous fish habitat upstream of a dam and severely harm downstream habitat. It would eliminate the endemic strain of Middle Fork Eel summer steelhead, a federally designated "sensitive species."

At the lower end, ocean beaches would be affected. The Eel River is the chief source of sand for the stretch of beach northward to the mouth of Humboldt Bay and southward for thirteen miles. A large upstream diversion project could result in extensive beach and seacliff erosion by reducing the sand supply. The importance of the Eel River estuary to fish and wildlife is well recognized. Flow regulation and reduction would cause severe disturbances in the estuary, with harmful consequences for estuarine-dependent organisms such as salmon and steelhead.

In sum, water moving down river channels plays many valuable roles. The undiverted North Coast rivers are not wasting away to the sea—they are *working* their way to the sea.

What Is "Wastewater"?

The popular notion that rivers simply "waste away to the sea" is a dangerous myth. There are several meanings of "wastewater." For chemists and sanitary engineers it means water that is degraded by discarded chemicals or human wastes. Specialists in irrigation systems consider wastewater to be that which leaks away or evaporates between the outlet at the dam and the plants in the fields. Water-thirsty people consider that water not used by man himself, or his plants or his factories, is wasted.

We need to refine our terminology. Agribusiness people consider that water not held back by dams and not flowing into canals is "wasted water." That water cannot properly be termed wasted, however, because it is the water that maintains the fish and other aquatic life of the flowing stream and the estuary where the river undergoes its process of mixing with and ultimately reaching the sea. Sanitary engineers use the term "wastewater" for the tainted and polluted water that comes from sewage treatment plants and factory effluents. Because this water has become, in effect, an ore that may not be processed economically, they think it best simply to dump it into the ocean.

The truly wasted water is the water that escapes between the points of diversion and the roots of the plants because of leaky gates and valves, unlined canals, and excessive flood irrigation. In California alone this loss is about 40 percent of all the water diverted for irrigation. It disappears into the air or sinks into the ground (where it may become part of the groundwater). Then there is much wasted water in the home, especially by toilet flushing, that adds unnecessarily to the wastewater of sewage treatment plants. Several states and counties are now requiring more efficient toilets for home use. We also use much water in excess for crops that may grow in other parts of the country without irrigation. Although it is pleasant to drive along the great rows of sprinklers on a hot day in the Central Valley, such air conditioning is conspicuous waste.

All of these wastes are part of our habit of externalizing the cost of our agriculture and industry, and those costs include such environmental degradation as fouling bays and estuaries and reducing fisheries at the expense of the essence of rivers or making domestic water unusable—the city of Sacramento is currently suing rice growers on this basis. To deal realistically with problems of wastewater, we must change our premise from the utilitarian, mechanistic realm to a broader view in which biological factors—the living essence of streams—are considered uppermost.

Without rivers and brooks, the land as we know it cannot exist. For several years during his boyhood in Vermont, George Perkins Marsh—whose book *Man and Nature* was the first environmental impact report on what we are doing to our world—was not allowed to read because he had strained his eyes, so he could only observe the world around him. He said that during that time "the bubbling brook, the trees, the flowers, the wild animals were to me persons, not things." Some years later, in 1853 on the distant western side of the continent, Chief Seattle made his famous speech of surrender to

the ways of the white man, in which he said (according to the version
in Joseph Campbell's *Historical Atlas of World Mythology*):

The shining water that moves in the streams is not just water, but the
blood of our ancestors. . . . The water's murmur is the voice of my father's
father.

The Rivers are our brother. They quench our thirst. They carry our
canoes and feed our children. So you must give to the rivers the kindness
you would give to any brother.

Chapter Sixteen

Steelie
Paul McHugh

It took a long time to find them. I'd glimpsed the silvery torpedoes of the steelhead trout—those huge, sea-wandering rainbows—when they thrashed up the waters tumbling down the old fish ladder at Van Arsdale Dam on the Eel River. But the vision only whetted my thirst, stimulated my imagination. I'd heard so much about these magnificent game fish, how they struck a lure like lightning, and fought like tigers once hooked. And more significantly, how they fought their way home to spawn, leaping even further up the whitewater of California coastal streams than salmon. . . . And how their life cycle, when permitted to play out, revealed California mountain forest, cool stream, and vast sea as connected in one organic whole.

On the North Coast, a steelhead has the potent status of a totem animal—it stands for all that is best about all that's still wild here. So even a dramatic glimpse of them struggling upstream against the swollen waters of the Eel wasn't enough. Their ancestral home was what I wanted to see, the clean gravel shoals of the upper headwaters. The place where these fish spawned, and their forebears before them. Where the waters contain a unique chemical "thumbprint," a faint smell that persists in the river's outflow some two hundred miles downstream. A taste of home, which these fish had sensed, and recognized, and turned toward weeks earlier as they swam the ocean coast off Humboldt County—as we might recognize the strains of a song unheard since childhood.

Sensitive to chemicals at the subtle level of parts per billion, they had noted and recognized the waters of this river estuary from scores of them along the California shore. They had run the gauntlet of sea lions, harbor seals, and human anglers at the delta mouth. Then a cold winter storm had swept over the Coast Range, and the spate of water rushing downstream had sent another signal to electrify the nerves of these powerful fish. Now. The time was now. It was time to go home. The water would be deep enough to leap the chutes of the rapids, cold enough to ensure the survival of eggs after spawning.

I started my hike up Bucknell Creek late on an overcast day, not long after a major storm. Twists and turns of the canyon, walls of tangled brush and jumbled rock, meant that soon I had to hike in the frigid creek itself. Soon after that it fell dark, and I sought a campsite.

When I awoke the next morning, my old army surplus bag was coated with a quarter-inch of hoarfrost. My shoes and socks were frozen stiff. Laboriously, I collected dry twigs from the ends of oak branches to make a fire that would thaw and dry them enough to wear. But within an hour, I found myself hiking in the foot-numbing water again. Past streambanks lined with drifts of old snow. Up boulders upholstered with frozen moss. Along jams of slippery logs. But every once in a while, I would see a long silver flash in a pool that would tell me the fish were still here, still traveling upstream. Once I even saw an improbable leap, a soaring, gymnastic effort that carried a big fish up and over a cataract to flap once or twice on a shallow rock and then slip safely into a pool.

Part of me felt pulled along by the drive of the steelhead toward the headwaters, a bullish determination to succeed at all costs. It was as though a part of my mind and heart had identified with them, and I continued long after my legs lost sensation below the knee due to the cold. Following the way of this totem animal, I continued fighting upstream until finally I found myself clinging by my fingertips to a steep canyon wall with no possible way to move even an inch further ahead, and facing a terrifying fall onto the rocks of a rapid below if I even tried.

The steelhead were still moving upstream as, defeated, I made my way back to my car by sunset and drove home. And fell into a week-long fever from the overexposure and exhaustion, with my temperature sometimes spiking to a neuron-threatening hundred and seven degrees. When my housemates brought me in to the emergency room, blood tests revealed no infection by virus or bacterium. But perhaps modern medical science has no way to treat someone who has temporarily lost a piece of his soul to a fish.

Weary of the strange dreams and visions of sickness, I at length broke the fever by opening my bedroom window and letting the cool winds of the new storm wash over my body.

After my strength returned, I went back to the headwaters of the Eel River. And finally, after hiking up another creek where I could stay on the bank, I found them. It was a clear crisp day, the sort of winter morning when brassy sunshine comes shouting down through cerulean skies. As I came around a bend in the creek, there they were, metallic bodies flashing in the sunlight as they sparred and mated in the crystal shallows. The silvery torpedoes of the males, darting against each other, shoving competitors into jets of current where they would be washed downstream. The long large female, weighing twenty pounds or more, her body flashing rainbow markings and the bright red blotches that steelhead acquire after their return to fresh water. I watched her turn on her side to dig her redd in a shoal of gravel with powerful thrusts of her tail. The victorious male joined her then, and their bodies shivered in muscular spasm, side by side, as the eggs drifted down into the nest through a cloud of milt.

Ah, the passion and raw power and wild beauty of that! When people mention the glory of the wild earth, this is my image—that supreme moment steelhead achieve after the ordeal of their pilgrimage, touching in the cold, clear creek, bodies spattered by sparkling chains of sunlight, the whole scene framed by forest and snowy hills.

We learn something important from watching the steelhead. They can remind us that some of the same force driving them impels us. Though at times its power is half-strangled by the fear or doubt in our trepidacious human minds. Though at times that force is hyped and distorted by those trying to sell our wildness back to us without a clue as to its essential dignity.

Unlike the salmon—who ascend their home rivers just once, to spawn and die in an orgiastic finale that seems the closest thing in the animal world to Greek tragedy—steelhead trout embody a tougher optimism. A few will survive, rest briefly, then head back out to sea, and return another year, and then the year after that, to repeat the arduous challenge of the spawning cycle. If they can.

The presence of wild steelhead and salmon in California rivers is charming, magical, fascinating. But—if you're aware of it—contrasting that rich presence to their absence can have an even stronger impact.

In Trinity County I was privy to a scene that suggested the glories of the past. Jim Smith, a World War II vet and county supervisor, stood in a long, dry gulch of rounded stones near his home. In an emotional voice, he told me about the salmon and steelhead runs that he'd seen here when he came home after the war. Runs that were nearly extinguished when a huge federal dam and diversion project took most of the Trinity River's water away in the 1960s.

"From hundreds of yards away, you could hear a bedlam of splashing, like kids in a swimming pool. The river was a series of big deep pools and long stretches of gravel with just enormous redds of salmon. You could wade out into the middle, and there would be salmon spawning everywhere around you. If you stood still, you were like a tree or anything else: fish would drift into the eddy of your legs.

"There'd be a pack of steelhead kind of idling downstream of a redd. One would reach in and lure the male off into an attack, and the rest of the steelhead would gobble up the salmon eggs. How could there be reproduction with that much predation? Well, the reproduction was just that much heavier. But personnel in the agencies don't believe me when I describe this because they have no experience or frame of reference for it; it just doesn't make sense to them."

The most poignant part of Jim Smith's story to me was not the loss of this scene of intense biological beauty from his childhood—all of us who grew up in the backcountry in the twentieth century have similar stories to tell—but what he subsequently did about his loss. For the next two decades, Smith fought in the political arena for restoration of the Trinity River. He gave endless interviews and speeches, wrote letters, organized committees, criticized studies and the bureaucracies that spawned them, and found like-minded souls in various agencies with whom to ally himself in the struggle.

And today, increased water flows are being returned to that ravaged stream. New spawning riffles have been built. A federally financed sediment catchment dam has been approved to reverse some of the damage. And the heroic efforts of an underbuilt hatchery are finally bearing fruit. And it was here that I finally felt the excitement of a tug-of-war with a steelhead at the other end of the line. My 1986 small fish, part of a run of "half-pounders," was a far cry from the hefty eight-and-a-half-pounder that was Bing Crosby's first steelhead, taken on a fly from this same river in 1963. That was the year the big water diversion dam was completed and the fish runs began their dramatic slide toward oblivion.

But if my fish was small, it was still part of a much larger run, numerically, than the mere thirteen steelhead that showed up here to spawn in 1977. The efforts of Jim Smith and his friends and allies were starting to pay off.

Once there were steelhead spawning in coastal streams as far south as Mexico. Once nearly all the northern California streams were choked with vibrant salmon and steelhead runs. Twentieth-century civilization has attacked them in ways almost too varied and numerous to catalog: water diversions and dams; siltation from road-building and logging; overfishing with instream gillnets; and pollution of various sorts.

The only really heartening part of the picture is the way that humans have also sought to make amends for their depredations. I remember volunteering for a few days on a project to clear old logging debris from the Albion River. Every summer morning that year in 1976, young people would stumble down into the cold river water to breathe chainsaw fumes and wrestle with wood, mud, and steel chains just to remove obstacles and improve the chances of the river's remaining steelhead runs. That gritty Albion River restoration scene, like Jim Smith's ultimately productive jog on bureaucratic and political treadmills, has also been repeated in various river valleys up and down the coast.

Some of that immense effort has been doubtless due to selfish motivations. After all, life provides few thrills like the one that comes when a big wild steelhead inhales your lure and sends the unmistakable quiver of the presence of a powerful life back up your line. It's like the tingle that precedes a thunderbolt, because the very next thing which happens is that the surface of the water

Catch 'em and let 'em go. Catch-and-release fishing is encouraged as one
means of preserving steelhead stocks. (Herbert Joseph)

explodes and the fish tries to tear the rod out of your hands, or,
failing that, strip all the line from the reel in a screaming downriver
burst of speed. It is not given to us to touch very many wild beings
in that way. I am thinking now of a specific twelve-pound wild
steelhead on a northern river in 1987.

But I like to think that a lot of the work for the well-being of
salmon, steelhead, and rivers comes from a deeper motivation: the
understanding that steelhead are canaries. That's right, canaries.
You've heard this metaphor before, but it's certainly true enough to
bear repetition. In the old days, coal miners would bring a caged
canary down with them in the shaft. Because the birds were so
sensitive, miners knew that if the canary ever stopped singing and
fell to the base of its cage, it was a warning that a poisonous gas was
spreading, and it was time to drop their tools and get out.

In the same way, vibrant steelhead runs are a living testimony to
the health of the streams, the forests that surround them, and the
oceans that are their vagabond home. When we can no longer

maintain runs of wild steelhead, an important quality of life will be dying. And soon after that, life itself for us may become rather difficult. So we try to reverse our damage, and keep them alive in the hope that we ourselves may survive as a species.

In the course of writing this piece, I've taken a few chances and talked about some things that I've never spoken about to anyone before. I'd like to continue with one final stretch for a piece of truth.

Once I got into trouble with the editor of a publication I worked for because I ran a photo of a man getting ready to eat a raw tuna heart. The man had just fought and landed his first big albacore tuna. The tradition of this particular tuna boat was that the angler was then presented with the fish's raw heart and commanded to eat it. In the photo the man's grin looks a little tight, but he has a can of beer in his hand to wash it down, and it's clear he's getting ready to do the deed.

The editor's position was that this primal image was going to make most of the readership feel like throwing their breakfast right up onto their morning paper. I acknowledged he was probably right, but I was still glad I'd run it. Because, in my opinion, we are a little too far away from the primitive belief that we take on strength from the things we fight, and the things we eat. And because if we are too squeamish to know the taste of the heart of a fighting fish, maybe we'll wind up too squeamish to get down in the muck with a saw and jam dirt under our fingernails as we work all day to partially restore a ruined river. Maybe we'll wind up without the grit and persistence we'll need to plumb the depths of our convoluted bureaucracies and emerge on the other side with the prize of a coherent policy.

A coherent policy that allies us with the wild beauties of our environment; instead of one that compels us to destroy them.

Part Three

Restoration Efforts
Political-Legal Perspectives

The broad outlines of environmental restoration schemes are determined in the legal-political realm. In that realm, the merits of a particular set of facts or feelings don't necessarily make a difference. Success in politics means making government work in one's favor. That is basic politics, the common element of the six chapters of Part Three. A corollary: The concept of virtue is out of place in law or politics because virtue is an unreliable concept—right to one side is probably wrong to another. Indeed, as the late Randolph Collier, powerful state senator from Yreka, California, is quoted in Chapter 17, "There is no rule that you have to be fair here."

The California environmental movement, of which salmon and steelhead restoration is an integral part, exists in the context of state water politics. So far, the alliance of agriculture and the water development industry has won more political support than have environmental restoration interests. Pete Chadwick, a Department of Fish and Game policymaker, recently stated the implications bluntly: during the 1990s, he predicts, success of fishery restoration efforts will depend more upon good politics than good science.

The authors represented here, as veterans of governmental and quasi-governmental service, understand these facts and implications very well. For reasons discussed in preceding chapters, all are convinced that changes in water management policies must occur. But not one of them is optimistic about prospects for major immediate change. Although the day of giant state or federal water projects may be over, the political history of water development in California, with its solidly entrenched interests and political traditions, is still being written. The January 1989 decision of the State Water Resources Control Board to distance itself from staff recommendations regarding water quality issues in the Sacramento/San Joaquin Delta revealed intensified determination of water interests to fight to retain their advantage.

The perspectives in Part Three vary considerably. We begin in Chapter 17 with a political success story for the environmental movement. The water industry and compliant—captive at times—government agencies have long wished to remove funds from northern California's water bank and send them south to enrich farms and urban communities in water-deficient central and southern California. They realized part of this goal with completion of the Trinity Division of the Central Valley Project. Still eluding them is a 30 percent larger branch of that bank, the Eel River. Former state Senator Peter Behr, in an interview with Jim Tarbell, recalls the intricate political maneuvering that accompanied his efforts to place the Eel and several other streams in the California (and ultimately national) Wild and Scenic Rivers System. As a freshman senator, Behr blundered immediately onto the turf of the "father of California's freeway system," Randolph Collier. His account of that experience reveals, as a textbook never could, the bare-knuckles strategies and tactics that mark California water politics.

In Chapter 18, Stanley M. Barnes, water engineer and chairman of the California Water Commission, presents in conciliatory perspective the ongoing "fish folk/water folk" debate. His 1989 speech at the annual California conference of the American Fisheries Society reflects a relatively new trend: the society, given to scholarly reports on technical matters, has in recent years incorporated presentations by legal and political decision makers in their conferences. Barnes, keynote speaker in a cluster of presentations by state and federal policymakers, describes what he perceives to be a changed climate for dialogue and assures a skeptical audience that a reasoned, systematic approach toward identification and solution of common problems would best serve all parties' interests. Particularly notable is his account of political efforts to ensure funding of improvements at Coleman National Fish Hatchery—still an unresolved issue.

The next four chapters deal with other legal-political perspectives of fish and wildlife restoration, as well as legal avenues available to help activists make government work in their interests. Chapter 19 focuses on a major concern: fish hatcheries. To most people in water-conscious California, artificial hatching and rearing of juvenile salmonids would seem to be the logical way to restore the fisheries. This belief fails to recognize that hatcheries require enormous amounts of cool, clean water. It also ignores the dangers of "hatchery dependency," a major concern of fishery professionals. Drawing upon genetic findings such as those reported earlier by Patrick Higgins, and viewing such data from the broad perspective of fishery management, Bill Kier explains why hatchery success can pose serious risks to natural stocks: natural runs falter as hatchery stocks flourish, with a resultant narrowing of the gene pool that makes all fish in a given population vulnerable to disease outbreaks and other natural catastrophes from which they may be unable to recover. Hatcheries have a respected place as a fishery management tool, but the need to maintain genetically diverse wild stocks as well is abundantly clear. Toward that end, federal and state legislatures have adopted policies encouraging habitat restoration to protect and enhance wild stocks and limit further growth of hatchery dependence.

The authors of the next three chapters discuss implications of the public trust doctrine, upon which much of the environmental movement relies. In his lucid discussion of this concept in Chapter 20, Felix Smith defines public trust, in part, as "the affirmation of the duty of the state to protect the people's common heritage of [water resources], surrendering that right of protection only in rare cases." This concept, its roots in Roman history, evolved through English common law, finally reaching its present form in American democracy. It is relevant to preservation of all natural resources, including the fish and wildlife that depend on the state's water resources. The CVP's effects on fish and wildlife resources provide the setting for Smith's presentation.

In Chapter 21, William Davoren invokes the public trust concept in his "Century of the Fish" speech to members of the California Senate's Committee on Agriculture and Water. The speech is built around a capsule history of water development in California. He dates the "Century of the Farm" from the late 1880s to 1982, when legislation authorizing construction of the Peripheral Canal was overturned by voters. Success of the new "Century of the Fish"—symbol for the remaining annually renewable values of our river systems—depends heavily on the outcome of the Bay/Delta controversy over water quality presently occupying the State Water Resources Control Board.

A more strident note marks Chapter 22. "Lock arms and say, 'I'm damn mad, and I'm not going to take it any more!' " That is the exhortation of William Sweeney, retired California area manager, U.S. Fish and Wildlife Service. This chapter presents one of the rousing lectures he began making soon after he left office, when James Watts was secretary of interior. In it he excoriates agribusiness and government agencies he holds responsible for the demise of Central Valley fish and wildlife resources. He explains how the public trust doctrine, migratory bird treaties, and the Endangered Species Act may be used as bases for legal action. Sweeney, like Behr with his coalition building, emphasizes that the success of environmental restoration efforts ultimately depends on the determined actions of an aroused citizenry working individually or through environmental organizations. The closing words of that speech: "How it all comes out is up to you." (In an epilogue dated August 1989, Sweeney points out that nothing has happened since he gave that speech to invalidate his advice.)

Chapter Seventeen

The North Coast Water War

An Interview with Peter Behr

Jim Tarbell

In 1969, before I got to the Senate, there was interest in North Coast water going to southern California, and the Corps of Engineers had settled on the Dos Rios area near Covelo as the most desirable location for a dam. It was to be a joint federal/state project, but Governor Reagan withdrew state support from the Dos Rios Dam. I think this decision was due to a Round Valley rancher named Richard Wilson who was very close to Ike Livermore, the director of the State Resources Agency at the time. I guess Richard talked to Ike, and Ike got the rest of the cabinet to talk to the governor, and he decided that it wasn't a good idea.

If that dam had been built, the water would have been moved over to the Sacramento River. It was a tremendously expensive project, and it would have taken an enormous acreage. At any rate, that was, and is, Reagan's finest hour as an environmentalist.

At that time they were already taking water out of the headwaters of the Trinity River at Lewiston Dam. The Trinity never has recovered from that dam. During the hearings on SB 107, the chairwoman of the Trinity Board of Supervisors came to Sacramento to testify that Lewiston Dam had ruined the Trinity and they hadn't got any jobs out

This interview first appeared in *Ridge Review* (July 1988); reprinted courtesy Jim and Judy Tarbell, publishers. After serving as a planning commissioner, Mill Valley city councilman, and Marin County supervisor, Peter H. Behr was elected to the California State Senate in 1970.

of it either. There were about forty different large dams suggested for all of these northern rivers. The Dos Rios Dam was the dam of choice, though, and when they lost that, it set them back a bit.

So when I got to the legislature in 1970, I introduced the California Wild and Scenic Rivers Bill as SB 107. I just put this bill into the hopper without ever discussing it with Randy Collier.

The Foe

Randolph Collier was the dean of the California State Senate and chairman of the Finance Committee, representing the North Coast for some twenty-odd years before I got there. He was from Yreka and was known as the Father of the Freeway System because, indeed, he was just that. He was a very wily and powerful old gentleman who had been known to say that "there is no rule that you have to be fair here." He knew all the tricks. He was tough. He was perhaps the most powerful state senator in the history of California and through whose district all of these rivers ran. He was simply outraged that I had introduced this bill, and I don't blame him. I quickly realized that I had made a terrible mistake and apologized. I told him I would never do it again, but I wasn't going to take it back. He was absolutely determined to beat this bill.

The Coalition

The impetus for the California Wild and Scenic Rivers Act came from different sources. I had been an environmentalist most of my life, but the spiritual father of this effort was a nationally known sportfisherman named Joseph Paul, who was born and raised in Eureka. At the time he was living in San Francisco and had started the Committee of Two Million (the number of sportfishing licenses issued in California each year). He was a man of tremendous public relations abilities and social skills, and in this committee he had collected all the sportfishing organizations in the state. He helped tremendously.

The Bill

A year earlier a California congressman, Jerry Waldie, had introduced a bill into the U.S. Congress declaring the Eel, Trinity, and

Klamath rivers as rivers under the National Wild and Scenic Rivers Act. I wrote Congressman Waldie and asked him if it would be a problem for him if I introduced the California Wild and Scenic Rivers Act in the California State Senate. He wrote back saying, "Go ahead, because this bill of mine isn't going anywhere."

So I took the descriptions he had of those three rivers and lifted them bodily from his bill and laid them into the California Wild and Scenic Rivers Bill. Then I patterned the rest of it after the National Wild and Scenic Rivers Act. The difference between the national and the state act was that, under federal legislation, if the act resulted from an act of Congress the federal government would have powers to condemn property along rivers far greater than the state's. In addition, the people at the federal level could control the Forest Service, the Bureau of Land Management, and the Corps of Engineers over building dams. The state would, or could, have trouble doing that.

The Battle Begins

Of these three rivers, the only one in serious contention for protection was the Eel River, which represented the plumbing of the North Coast. So the battle was over the Eel, and the first thing Randy did was introduce a bill asking for a study. If you wanted to slow things down, you asked for a study. His bill directed the secretary of the Resources Agency to prepare detailed waterway management plans for all nineteen rivers and streams in his First Senatorial District. Not just the Eel, the Klamath, and Trinity, but all the ones along the coast, including Big River, the Garcia, the Navarro, the Noyo, and so forth. There was no deadline and no moratorium in his bill while these studies were being prepared.

The coastal rivers were not included in my bill because we relied on a study that rated rivers by priority of national importance for scenic and other purposes. The Trinity, Klamath, and Eel all received the highest priority in this study, but the rivers along the coast did not.

Slowly we began to gather momentum for this bill. We had a strange coalition. What was really significant was that for the first time the fishermen joined with the conservationists, not as equals, but with fishermen leading the efforts. We had the Native Sons of the Golden West, the Native Daughters of the Golden West, the

Teamwork. This citizen volunteer and Department of Fish and Game biologist are gathering data on summer flow of the upper Eel River below the Van Arsdale diversion to the Russian River. (Herbert Joseph)

California Real Estate Association, and every conservationist group you could think of, including Friends of the River. We had all the California Rod and Gun Clubs and all the California Sportsmen's Associations. Then we started to knock off the counties. We got Trinity, Siskiyou (where the old gentleman had his home in Yreka), Marin, but not Mendocino or Sonoma.

Collier didn't think it was going to be much of a problem defeating the bill, because it was going through his Finance Committee, and he figured he could handle his own committee. In addition, it needed a two-thirds vote on the Senate floor.

Now, there was a senator named Lou Cusovanich from down south who was in Randy's camp. He was scared to death of the old man, always voted with him. But the Friends of the River raised such hell in his district that he came to me and said, "Look, these crazies are driving me up the wall and I'm going to have to vote for your . . . bill." That was Friday. We put it on the calendar for

Monday. Collier never questioned the fact that he had his vote. So I got it out of the Finance Committee with one extra vote. But then Collier did me in on the floor of the Senate. I couldn't get more than nineteen votes. Cusovanich was quickly brought in line. So I lost that, and that was the end of the year. Then the next session I put the bill in again. By then we had begun to get a great deal of momentum behind the bill. In fact, we thought it would be very difficult to stop. Then Joe Paul died. We realized then that we could lose the bill because he had been so important and everyone had loved him so much and he could scare the bejeezus out of Randy Collier. But we went right ahead in memory of Joe.

In that second year Randy Collier introduced SB 4, a carbon copy of my bill but excluding the Eel River. And every time we would amend my bill, he would amend his. I couldn't stop his bill, and he couldn't stop mine.

His bill was supported by the Metropolitan Water District of Southern California, all of the irrigation associations, the California Water Resources Association, and the Eel River Water Council with Jerry Boucherg. The counties along the Eel River created the Eel River Water Council to preserve the water of the Eel for those same counties, but it was captured by the major water interests who became members. Jerry Boucherg became the executive director and lobbied for the big Metropolitan Water District and everybody else. The whole wonderful world of water was opposed to SB 107.

We led a charmed life in the end, though. My bill was tied up because Randy Collier wouldn't set a time for it to be heard in his Finance Committee. We turned on the heat. We went to the Rules Committee and raised hell until Collier heard it, but then we only had three weeks until the end of the session. In three weeks it had to go through the floor of the Senate, through two Assembly committees, and off the floor of the Assembly; and we had to do it with no changes, or we would have had a conference committee on the Assembly side and we'd be screwed, blued, and tattooed.

So we did it on the last night of the Assembly, which is always a madhouse. Nobody knows what is going on. You can't. Things are going too fast.

On this last night, the speaker, Leo McCarthy, whom I knew and worked with, put the bill up for a vote, and when the Speaker puts a bill up for a vote, it goes through because he has the ultimate ability to punish you. So Leo took it off to the governor on the last night of the Assembly. Both bills landed on Reagan's desk at the same time, and Reagan had to choose which to sign.

We had to draw up very careful charts showing why and how the Collier bill differed from mine. By this time Collier's bill had all sorts of little loops and crannies designed to make the bill harmless to the big California water interests that sponsored it. These loopholes were very hard to find. He was a very agile man. But we finally got the governor to sign my bill.

On to the Feds

I had put a stipulation into my bill that the secretary of the State Resources Agency request the secretary of the interior to place the California wild and scenic rivers in the National Wild and Scenic Rivers Act.

Then years went by, and what happened was amusing. Near the end of Governor Jerry Brown's term, Huey Johnson, secretary of the State Resources Agency, suddenly realized he was the secretary of the Resources Agency and could request that these rivers be placed into the national system. So he called up the assistant secretary of interior and there was no problem except that there had to be an environmental impact report (EIR).

So the EIR was going to cost $350,000. The federal government said they didn't have any money, and Huey didn't have any money. So Huey decided he would tithe all the departments in his agency to come up with this $350,000. The problem was that you can't move money around without approval of the legislature. But the Budget Committee never found out about this until later, and then they were absolutely outraged. They were going to strip him of his medals and throw him out. But anyway, it became part of the national system.

The Future

Though Congress can do it, nothing has ever been taken out of federal wilderness protection. When Doug Bosco was in the California Assembly, though, he did take some of the Smith River tributaries out of the Wild and Scenic designation. Initially I had not included the Smith River in the bill because it wasn't really threatened. To make up for it, I included all the tributaries to the Smith. So when Bosco was an assemblyman, he took out about sixty miles of the tributaries to benefit a mining operation.

But I think, in general, the rivers are pretty safe for future generations. It is my understanding that the upper echelons of both the Corps of Engineers and the Bureau of Land Management have decided that there will be no more large dams. Large dams cost too much now; their cost overruns are very high, and most of the good locations have already been taken. They are similar to atomic plants; no one wants them any more. Then, of course, they have to get appropriations every year, and if you start fighting it, and it takes eight or ten years to finish a dam, you can usually kill it by then.

No, I don't think the North Coast rivers are at risk but, then again, if it involves water, they are always at risk.

Chapter Eighteen

Water and Salmon Management in the 1990s

Stanley M. Barnes

I was quite interested last night listening to Allen Johnson. I asked some of the folks sitting around me, "Would that kind of speech have been made ten years ago?" and the reply was, "Probably not." On the other hand, neither Bob Potter nor Larry Hancock nor I would have been asked to show up for a meeting of the American Fisheries Society; and if we had been invited I don't know what we would have said that would have been constructive. Looking out at Don Kelly [former CDFG official], I think we first got acquainted following a time of great disharmony between the water folks and the fish folks. The fish folks were putting a little pressure on the new director of the Department of Water Resources, Dave Kennedy, and the water folks were putting a little pressure on the same fellow. Now, Dave isn't the dumbest guy in Sacramento, you know, and he asked the fish folks if they would be willing to talk to the water folks and the water folks if they'd be willing to talk to the fish folks. We all agreed to do that. The first meeting that I remember took place in Los Angeles, in a not-too-grand meeting room, but at least there was a table, and we all sat around it. My recollection of the meeting was that only two things happened. First, one by one, we all insulted everybody on the other side of the table, and received a reciprocal insult in one form or another coming back at us.

From the keynote address at the annual conference of the American Fisheries Society, Napa, California, February 11, 1989.

Second, and the only constructive thing to come out of the meeting, was that we agreed to meet again.

From that inauspicious beginning has come the two-agency fish agreement, which nobody thinks is too terrific, but it was a positive step. So I think we are heading in good directions.

I might give you very briefly just a summary of who our people are on the California Water Commission. There are nine of us appointed by the governor, confirmed by the Senate, four-year terms. I've been on six years. Our fishery enhancement committee consists of Clair Hill from Redding, Audrey Tennis from Chico, and myself. The other members of the commission include Jim Lenihan from Santa Clara Valley, Jack Thomson, a farmer from Bakersfield, Martin Matich from San Bernardino, Harold Ball from San Diego County, Katherine Dunlap from Los Angeles. Our other member, Lee Henry, has just resigned to go to Great Britain for an eighteen months' stint for the Mormon church. So we're without a person. The group really does look at itself as trying to do something to provide for and participate with others in constructive solutions to all these situations that we've collectively allowed to deteriorate, I think, relative to the fish problem.

Among its other activities, our fishery enhancement committee has been involved as an informal monitor, you might say, of the Upper Sacramento River Fisheries Advisory Council. We've been very favorably impressed by the fact that there has been participation by a broad group of federal, state, and local interests, including the farmer landowners along the river. Many of the participants have come to realize that it makes more sense to work out problems amicably than to fight continually against things they don't really oppose. There really are many more common objectives on fishery issues than they had believed. We look forward to seeing the report. I don't know of anyone who thinks it's the last word on Sacramento River problems. But the council did look at twenty fishery items and two riparian items. They've put priorities on them, and they've put estimated costs on them, and have said, "Hey, let's look at these matters in the priorities listed."

I'd like to talk about two things that I think collectively we can have some influence in—and when I say "we," this isn't just the Water Commission. These are matters on which the water and fishery communities can have some impact. One of them is Coleman National Fish Hatchery. It has disease problems; it has water temperature problems; both are well known to the Fish and Wildlife Service folks. They're as frustrated as any of the rest of us. They have an expensive, carefully developed plan for rehabilitation of Coleman that totals some $22 million. The proposal is to do approximately $7 million worth of work in Phase One, which would be water treatment, ozonization of the waters for disease problems, and cooling. The second phase would be approximately another $7 million, which would be for additional water treatment and cooling. The remaining money would be for additional water treatment, but would not be spent except as needed and as the results of the first two phases became known.

That report was prepared in draft form two years ago; it was finalized in November 1987. The California Water Commission, through its appropriations process, has been meeting with the technical people and the regional director types from the Fish and Wildlife Service and the folks in Washington. The situation we find is that not only can we not get the $22 million, we can't get the $14 million, we can't even get the first $7 million. We can't get a penny! The local folks at the Portland level have put in requests for the president's budget in amounts for at least a start on the program, at least a few improvements. However, the only things that have been done in the way of funding have been congressional add-ons. The administration simply, for "budgetary reasons," has not seen fit to fund these things. We sat in a meeting in Washington—I don't want to get too far afield, but I want you to know the perspective from which we see the thing—we sat in a meeting with a gentleman named Bill Horne, who was assistant secretary of the interior for fish and wildlife and parks. He's, you might say, the boss of the Fish and Wildlife Service, and we said, "We're here to try to get this thing funded." He replied, "My national budget for new construction of fisheries facilities in the United States is $15 million, and you don't fit into it." So we said, "Suppose we just get a little congressional add-on. Give us a line item and we'll get a congressional add-on." "No," he said, "if you do get the funds, I'm committed to the administration to see that they're not spent." Now, when the highest level of the Fish and Wildlife Service looks you in the eye and says those things, ladies and gentlemen, you've got problems.

This last fall the commission made a serious run at funding Coleman improvements—the Portland office put in for $1.8 million total in three separate specific line items for Coleman rehab; it came back from the president's budget—I've not seen the Bush budget, but the Reagan budget was for zero. Zero in each item. It's an impossible situation. In the meantime we, as a commission, have worked with the California state administration and with the Central Valley water users and the Central Valley power users, actually through WAPA, and we got an agreement by those folks, a commitment from Secretary Van Vleck and the directors of the Department of Fish and Game and the Department of Water Resources, to have state funding for Coleman rehab at the same level as the Trinity formula, which most of you know is 15 percent California. This would be a million dollars of California state money for Phase One, another million for Phase Two, and the water and power users agreed that they would pay, as reimbursement from water tolls and power charges, that they would reimburse 50 percent of the total. So, from our perspective, the local people have done all that you could ask of them. The water and power users have said, yes, we will cooperate, we'll pledge our funds, and the state has said they will do that. But some place we've got to adjust things at the federal level. The proposal now is that we would seek to have the custody of Coleman returned to the Bureau of Reclamation. Funds would then be provided through congressional committees that are more sympathetic to the problem, and more sympathetic to correcting the problem, than the committees with present jurisdiction appear to be.

This is not in any way meant to be a slap at the technical people or the operational people at the Fish and Wildlife Service, all of whom we feel have done as well as they could without money—but how it will shake out, whether it will end up like Nimbus, owned by the United States, administered by the Bureau of Reclamation, and operated by Cal Fish and Game, whoever it would be, or just funded by the bureau and the money go back into Fish and Game, that hasn't been determined. But anyway, we are working on that, and we think it's an important program.

Another program that I want to comment on is a matter that a lot of people in the audience know a great deal more about than I do, so I hope you'll forgive me. I'm talking about the spawning gravel restoration program in the Sacramento River system. This is a matter on which the commission—and not just the commission but individual members of the commission, and members of the water community, and the Departments of Fish and Game and Water Resources, and some of the federal folks—can have an impact. There is a compounding of jurisdictions of different state agencies, all well intended, and if you take each regulation by itself, it may make a great deal of sense. But when you put them all together, you've got an almost impossible situation.

The story unfolds about like this: You have a federal agency in charge of handling releases from Shasta and Keswick, and they can't just turn the valve on and off whenever they want to. They have certain water requirements both for their users through the Sacramento and San Joaquin systems and also for Delta outflow requirements, so they don't have a completely free hand. They're operating within the Coordinated Operation Agreement and every-thing else that most of you are quite familiar with. Next, there is an agency called the State Reclamation Board, which is concerned with flooding, making sure that these floodways, the Sacramento River and major tributaries to the Sacramento River, have the capabilities to take floodwaters away without shoving the water to the other side of the bank on the neighbor or something like that. Because of that, during the flood season, from approximately the middle of October through perhaps May, you can't stockpile gravel within the floodway. You have to stockpile the gravel outside the floodway. Where do you stockpile it? That's the problem. Taken by itself that's a reasonable requirement.

Another requirement is from the Central Valley Regional Water Quality Control Board. They don't want silt in the water. The fish folks don't want silt in the water. But someplace there's got to be a reasonable balance—you can't take the spawning gravels mined from someplace and wash them to the point that when you put them in the stream it causes absolutely no silt whatsoever. A good half-inch or an inch of rain in the same watershed would create a great deal more turbidity. So someplace or other we need to find a balance between what is reasonable and what is not, relative to the turbidity for the short period of time when you would be placing the spawning gravels.

Fish and Game has its own regulations relative to what is reason-ably required relative to the fish and when you can put these

Chinook salmon in spawning grounds of a Sacramento River tributary. (Dave Vogel)

gravels in relative to spawning time and so on. Someplace along the line we hope to get all of these agencies together in the same room and chat about a program where you don't have to pay three to five times as much for the gravel as you would otherwise pay, and where you can find some way to stockpile reasonable quantities of it so you can achieve the common objective, which is to get the fish counts back up. That's really the objective. Someplace or other when we're establishing these criteria, whether we be the Department of Fish and Game or the water quality control board or whoever, we want to look at the final objective. The final objective is to make the doggone thing work for the fish.

Another element that's somewhat related to that is the increased pressures for mining of gravel for commercial gravel operations. The one that strikes me is the one recently on Cottonwood Creek between Tehama and Shasta counties. That got caught in local politics, and some of us took the view that it is really a more important issue than just a zoning thing or just a permit process through a local county. To spend tremendous amounts of time, energy, and money to restore spawning gravels at one location in the system and have someone else destroying them at another location is probably not in the best interests of the whole program!

The other message that I'd just like to touch on briefly relates to fishery impacts of the State Water Project. Most of you know that I represent people in my engineering practice who are users of water from the State Water Project. There's no denying that the State Water Project has had some adverse impact on the fisheries in the Delta system. But our message goes like this: the problems are many and varied; the causes are many and varied; the State Water Project and the federal Central Valley Project have caused some of the problems but not all of the problems. There are other water projects that have caused problems. What we need to do is continue to work as we have in recent months, and for the past couple of years now, in trying to develop specific solutions for mitigation of the Delta pumps, mitigation of the State Water Project operations, but also to look at the system as a whole, to look at how can we really achieve mitigation of the whole system. I'm not trying to point the finger at Hetch Hetchy, East Bay MUD, or anyone like that. I'm just trying to say, "Let's take the system as it exists, and look for specific solutions. Let's address those potential solutions, and not spend too much time trying to find out who caused the problems."

I heard a pretty good statement at a recent meeting, which went about like this: Here's the problem, do we have a program or programs for its solution or at least its partial solution—some positive steps? Yes we do. We have program "A" or program "B." Okay. Program "A" will be a positive thing; it may be partially mitigating or help to resolve problems caused by a dozen different people or agencies or activities. Let's do this first program, and let's find a reasonable cost allocation, and go about it, and seek funding from a variety of sources to do it. My experience is that when you come back to the people, whoever they may be, water people or fish people or anybody, when you come to them with a specific program that makes some sense, both physically and economically, you can get cooperation in terms of both personal participation and money.

Chapter Nineteen

California Hatcheries

They've Gone About as Fer as They Can Go!

William M. Kier

Between 1872 and the 1920s, California fishery managers were enamored with the idea that depleted stocks could be restored artificially. Hatcheries (first called "fish breederies") seemed to hold great promise. California has operated as many as one hundred and sixty-nine artificial reproduction facilities, ranging from simple egg-taking stations to factory-size hatcheries handling millions of eggs and fry.

Fish of many species were planted willy-nilly in about every stream and lake in the state. Resort developers arranged for hatcheries to promote tourism. Politicians promoted hatcheries for their home districts.

Over the years, enthusiasm for hatcheries has waned. Today there are nine government salmon and steelhead hatcheries in California. Eight of these were built by dam developers under laws that required them to "mitigate" for blocking upstream spawning migrations. (The ninth, the Mad River Hatchery, was originally built by the state to "enhance," or increase, salmon production.)

Although the art of fish culture has become quite sophisticated, problems apparent almost from the start persist. Other states and countries have experienced serious difficulties with artificial fish production programs, and now California has set the stage for its own costly salmon and steelhead hatchery disaster.

Projects to exploit the streams of California's Central Valley,

primarily for irrigated agriculture, have cut off salmon and steelhead trout from 95 percent of their traditional spawning and nursery habitat. As the scant remaining stream habitat has been steadily
degraded by water project operations, the valley's remaining
spawning runs—70 percent of the state's total—are increasingly the
progeny of fish that were artificially spawned at the region's mitigation hatcheries. In this way, California has become unwittingly
"hatchery dependent." This dependency has worried some California fishery scientists and managers for more than a decade.

The survival of salmon and steelhead through the ages was possible because of the genetic diversity created through the interaction
of natural breeding populations. As drought or disease diminished
one race of fish, another race, better able to adapt to the particular
hardship, would take its place. Wild populations with sufficient
genetic diversity show amazing resilience to natural environmental
change.

By their very nature, modern hatcheries, like modern corn-
fields, create a lot of progeny from very few parents and, consequently, narrow the total "gene pool" over time. In contrast, natural, inriver spawning maintains and enlarges salmon and steelhead
genetic resources by distributing the fish over the available spawning grounds and throughout the months of suitable spawning conditions. Natural spawning therefore spreads the risks that a population of salmon or steelhead can be decimated by drought, disease,
or other disaster, while hatchery spawning, restricted in time,
space, and genetic variability, heightens those same risks.

Recent studies of the steelhead trout in Oregon's Kalama River
watershed found that the offspring of wild trout there survived at
three times the rate of those produced by hatchery-born spawners.
Oregon's effort to increase silver salmon production by expanding
hatchery output, beginning in the early 1960s, led to a gradual
deterioration of the survival in the wild of *both* hatchery- and
stream-produced fish.

Canada began a costly expansion of king salmon hatchery production in the 1970s. The salmon catch in Canada's Georgia Straits
collapsed in 1987—down 80 percent in less than a decade. Scientists believe the increasingly poor survival of hatchery fish led to
Canada's costly mishap.

California hatchery managers have increased the efficiency of their Central Valley facilities enormously since 1968 by growing larger fish, at substantially higher holding costs, and trucking the fish to the San Francisco Bay-Estuary—or even releasing them directly into the bay. At the State Water Resources Control Board's 1987 Bay/Delta water rights hearings, experts reported that, as a result of these improved hatchery operations, 46 percent of the king salmon returning to the Sacramento River system to spawn now come from two artificial production facilities: the Nimbus (American River) and Feather River state hatcheries.

During the same period that the hatcheries have become so successful, the Sacramento River's natural salmon and steelhead spawning populations have deteriorated steadily. What were once four healthy races of wild king salmon spawners—fall run, late fall run, winter run, and spring run—have dwindled to a fall run half its former size and two remnant runs; one, the winter run, is precariously close to extinction.

California's hatchery dependency—still viewed by the state's water development community as the final solution to the conflict between water exploitation and fish conservation—has set the stage for serious trouble. A disease outbreak, an event that could be tolerated by the region's genetically diverse wild breeding populations, could literally wipe out California's Central Valley salmon and steelhead resources.

California's fishery professionals have understood for a long time that the protection and improvement of stream habitat is the only responsible means of assuring the conservation of California's salmon and steelhead trout resources—and they have appeared powerless to stop the constant assault against stream habitat by dam builders and other developers. Recently, however, their concern has found its way into state and federal policymaking. In recommending adoption of 1986 legislation to restore the fishery resources of California's Klamath River basin, for example, Congress's Committee on Merchant Marine and Fisheries commented: "The Committee wishes to emphasize its view that rehabilitation efforts to support restoration of wild stocks offer the most significant and cost-efficient long-term benefits and should be the primary focus of the program."

Two years later, in adopting the Salmon, Steelhead Trout, and Anadromous Fisheries Program Act to conserve and restore the

state's dwindling anadromous salmonid stocks, the California legislature commented: "Reliance upon hatchery production of salmon
and steelhead trout in California is at or near the maximum percentage that it should occupy in the mix of natural and artificial hatchery production in the state."

As Rodgers and Hammerstein noted of turn-of-the-century Kan-
sas City, "They've gone about as fer as they can go!"

Chapter Twenty

Water and Salmon Management in the Central Valley
Felix E. Smith

The state of California is the trustee of its waters. The state, as trustee, is empowered to bring suit to protect the corpus of the trust—the water—for the beneficiaries of the trust—the people. The state is also the trustee for the fish and wildlife resources found within its borders.

Water is free in California. There are no royalty charges. When one receives a water right permit or license, one pays an application fee and the cost to develop and transport the water. Water is seldom free of public trust interests. It is unconstitutional to waste or destroy water. When the water returns to natural watercourses or other waters of the state after use, it must meet water quality standards to protect the public and private interests and trust uses of those waters.

Water is both a resource and an ecosystem. All too frequently, however, water is viewed as a commodity. Water development is the basis of California's prosperity. California's natural, renewable resources have been altered drastically to achieve California's modern prosperity.

This essay is taken from the introduction of the June 1987 discussion paper "Water Development and Salmon Management in the Central Valley of California and the Public Trust"; revised August 1989. The views expressed do not necessarily represent those of the U.S. Fish and Wildlife Service or the Department of Interior.

Water as a commodity is transferred from one region of the state to another. As a commodity, water is used as a raw material in agriculture, in industry, and in the home. When this water has been used, or when it is no longer needed, or when it has been severely degraded, it is cast aside or thrown away. This casting aside usually occurs as a point-source or a non-point-source discharge to surface waters or, in a few instances, to the groundwater. In this process the role and function of water, as an important life support system, as an ecosystem important to fish and wildlife resources and for numerous other beneficial uses, are usually ignored.

Social and political values once supported a water management system that destroyed entire rivers in order to serve agriculture and cities. Friant Dam–Millerton Reservoir (Bureau of Reclamation) on the San Joaquin River is probably the most dramatic example in central California. In northern California the Trinity Project (Bureau of Reclamation) has become infamous for its impact on anadromous fish resources. About 90 percent or more of the Trinity River runoff was diverted at Lewiston to the Central Valley for agricultural purposes. In the last few years diversion to the Central Valley has been reduced and greater streamflows have been released to the Trinity River as part of a special study. There has been an increase in the number of salmon and steelhead returning to the spawning areas. A portion of this increase is attributed to greater instream releases. People have forgotten or are too young to know about the great runs of chinook salmon that used to migrate up the San Joaquin River and spawn where Millerton Reservoir is located.

Today there are indications that a more balanced water management policy is being pursued. Changing social values and great public pressure to protect the public rights, interests, and resources of small streams in faraway places like Rush Creek and Mono Lake, as well as well-known places like the Sacramento/San Joaquin Delta and San Francisco Bay-Estuary, are gaining public attention.

Change Is in the Wind

In the Mono Lake case (*National Audubon Society v. Superior Court*, 33 Cal. 3d 419; 189 Cal Rpt 346, 658 P.2d 709, February

1983), the public trust doctrine was the primary legal basis used by Audubon. The meaning of the Mono Lake decision is:

1. The state has an affirmative duty to take the public trust uses into account in allocating water and, insofar as feasible, avoid harm or degradation to public trust resources, uses, and values.
2. The state has continuous jurisdiction over previously issued water rights and therefore can reconsider previous water allocation at any time.
3. The public trust is an affirmation of the state's duty to protect the people's common heritage of streams, lakes, marshlands, and tidelands, surrendering that right of protection only in rare cases when the abandonment of that right is consistent with the purposes of the trust.
4. The public trust doctrine, under which the state holds title to navigable waterways and the lands lying beneath them as trustee for the benefit of the people, protects navigable waters from harm caused by diversion of nonnavigable tributaries.
5. The public trust includes the protection of ecological and biological values.
6. Any member of the general public has standing to raise a claim of harm to the public trust.

In the Delta Decision (*United States v. State Water Resources Control Board*, 227 Cal Rpt 161-1986), the court ruled or reaffirmed among other things that:

1. Water rights are limited and uncertain.
2. The rule of reasonable use is now the cardinal principle of California water law.
3. The State Board has a mandate under state and federal laws to set water quality standards necessary to protect fish and wildlife.
4. The State Board has authority to modify permit terms and conditions to prevent waste or unreasonable use or unreasonable methods of water diversion.
5. No party has a vested right to use water in a manner harmful to the interests protected by the public trust.

These and other points, in combination with the Mono Lake Decision, set the stage for future water management consistent with the public trust and in the long-term public interest.

The Appellate Court decision in *California Trout v. State Water Resources Control Board* (207 Cal. App. 3d 585-1989), a case involving diversion dams on four tributaries to Mono Lake, supported the understanding that fish as a trust resource have a unique status. The title to and property in fish within the waters of the state are vested in the state and held by it in trust for the people.

The Appellate Court also indicated that the provisions of Section 5937 of the Fish and Game Code—"The owner of any dam shall allow sufficient water at all times to pass through a fishway, or in the absence of a fishway, allow sufficient water to pass over, around or through the dam, to keep in good condition any fish that may be planted or exist below the dam"—are an expression of both the California constitution and the state legislature for protecting the value of the state's instream waters as an ecosystem and the fish resources that use that ecosystem. The effect of that provision is to limit the amount of water that may be appropriated by diversion by requiring that sufficient water *first* be released to assure that fish life below the dam is maintained in good condition.

The criterion "in good condition" is not defined in Section 5937. However, "in good condition" should include the conservation and protection of the biological, physical, and chemical aspects of the aquatic environment that are necessary to support self-maintaining or renewable fish populations, associated ecological values, and other beneficial and public trust uses of the stream. The Appellate court also stated that the public interest of a fishery in a non-navigable stream is in the nature of a property interest and that there are a variety of public interests in addition to fish and the fishery that pertain to such waters.

The State Water Resources Control Board (State Board) and the regional water quality control boards collectively are responsible for protecting instream flows and the quality of all waters of the state. These boards have a continuous responsibility to review, amend, or revoke any permit or license in order to protect water quality, public health, fish and wildlife resources, and other aspects of water having intense public interest. This applies equally to a water right and to a discharge permit.

Historical Summary

California's Central Valley is a continuous valley extending over four hundred miles from Red Bluff in the north to Bakersfield and adjacent foothills in the south. This valley is bordered on the east by the Sierra Nevada and on the west by the Coast Range. The Sacramento River and tributaries flowing in from the north and the San Joaquin River and tributaries flowing in from the south form the Delta as these waters merge and flow to San Franciso Bay, the Golden Gate, and the Pacific Ocean.

Before the construction of levees and reservoirs on the major Central Valley rivers, rainfall and snowmelt runoff during the fall, winter, and spring months frequently flooded the lowlands. This flooding provided several important ecological functions. It helped cleanse the stream and river channels, marshes, and wetlands of any natural silt, salts, and massive amounts of decaying organic material. This material, washed downstream to the Delta and San Francisco Bay, helped nourish and renew the estuarine ecosystem. The flooding process also irrigated the valley's marshes and grasslands with fresh water that supports numerous migratory birds and other wildlife. This flooding cycle, repeated annually, was a critical component of the ecological process. The gravels used by chinook salmon and steelhead for spawning were also cleansed of their silt and replenished. These runoff waters also helped young salmon and steelhead migrate to the Delta/San Francisco Bay and the Pacific Ocean.

In the early days of California, the process for obtaining water was a simple one. Rivers were dammed and their waters diverted to meet local economic needs. The riparian owners diverted water for mining, agriculture, and household uses. As cities grew and agricultural demands increased, the rivers nearest the area of demand were developed and diverted first. The large rivers at a distance went next and so on. While this was occurring, there was little concern for instream uses, products, and values. The thinking and reasoning at the time was simplistic: there is little rain in the Central Valley during the summer and the dry lands needed water to produce crops to feed a growing nation. In this vast state there seemed to be an endless supply of rivers from which the desired water could be obtained, an abundance of marshlands to reclaim, and vast areas of good land to irrigate.

At the time little concern was expressed for protecting the enormous runs of salmon and steelhead that migrated up the great rivers. Just as there seemed to be an endless supply of rivers to supply water, this same endless supply of rivers would be available to meet the needs of salmon, steelhead, and other fishes. These fishes, it was assumed, would use one of the other rivers for spawning and rearing. In addition, the wetlands were deemed wastelands to be diked, drained, or filled. Vast acreages were diked and drained for agriculture or for urban and industrial growth. The economic values, products, and opportunities (both commercial and recreational) provided to communities by these fish and migratory bird resources were usually far removed from, and out of sight of, the community that diverted the water.

Any public interest valuation at the time was limited to the traditional water uses and values such as irrigation and municipal and industrial supplies. These uses were more easily translated into monetary "beneficial" terms, especially at the local level, than were the public values of instream flows, water quality, renewable wildlife and fish resources, and aesthetic and other instream uses. As a result, instream flows, wetland and aquatic ecosystems, as well as fish and wildlife resources and their values to commercial fisheries, recreation, and tourism, did not fare well. In some cases streamflows were so altered that there was no remaining aquatic ecosystem and therefore no resource renewability. In a few streams, the entire flow was diverted and the streambed left virtually dry. In the San Joaquin River, for example, a large salmon run was eliminated, groundwater recharge was greatly reduced, wetlands were left dry, Delta inflows were eliminated, and the quality of any remaining water greatly degraded.

The Picture Today

Protection, conservation, and prudent use of the state's waters are currently matters of great public concern. The waters of the Sacramento and San Joaquin rivers, the Sacramento/San Joaquin Delta, and San Francisco Bay-Estuary are critically important habitats for chinook salmon, striped bass, and numerous other fish species and to the waterfowl and other migratory bird resources of the Pacific flyway. As the number and quality of our free-flowing streams decline and the number of people desiring instream resources grows, the total value of the remaining rivers increases tremendously. These rivers provide nursery and spawning areas for chinook salmon that support sport and commercial fisheries far removed from their parent stream. They are areas for recreation, open space, and transportation. They are valued as public places. These rivers and their associated wetlands, even in their degraded state, support runs of salmon and steelhead, millions of wintering waterfowl, and other migratory birds. These water and wetland ecosystems contribute significantly to our aesthetic sense, to our appreciation of natural beauty, to the economy, and to our daily lives.

In simplistic terms the conflict over water, its uses, products, and values, is based on the scarcity of water both as a commodity and as an ecosystem. But in an absolute sense there is no scarcity of water. What is scarce is water in both amounts and quality to assure the continued maintenance of instream flows, the renewability of resources, uses, products, and their associated commercial, agricultural, and recreational opportunities. There is a shortage of cheap or subsidized water to irrigate or reclaim vast areas of semidesert while at the same time assuring the renewability of instream resources and protecting numerous instream beneficial uses in the area of origin.

Municipalities and industries are capable of paying, and are paying, significantly higher prices for water compared to agriculture. In some areas, agriculture is competing against agriculture for water. There is private water as well as water from the State Water Project and the federal Central Valley Project. Products from this agriculture frequently compete in the marketplace. In other areas, industry is competing against agriculture. In some cases industry is reclaiming and reusing its wastewater at a price higher than it would be paying for new water. Urban and industrial users are expected to meet wastewater discharge standards, while agricultural users downplay the hazards or toxic effects of agricultural wastewater or chemicals in the groundwater while complaining about (or claiming exemption from) federal and state clean water standards.

The public is starting to focus on those who have taken more than their fair share of water and on those uses of water that may not be considered reasonable and beneficial. For example, there is

heightened concern regarding (1) the continued statewide movement of water, (2) the intense competition for water, (3) the need to
protect and assure the renewability of fish and wildlife resources,
(4) contaminants in the aquatic environment (including groundwater) and their potential impact on beneficial uses, (5) water conservation, and (6) the need to protect other interests, uses, or
values covered by the public trust.

As developed water becomes shorter in supply, as agricultural
chemicals and trace elements degrade water quality and affect the
beneficial uses of water, as the public learns more about fish and
wildlife resources lost and opportunities forgone and develops restoration skills, the past failure of government agencies to accept
and implement their public trust responsibilities for managing the
people's water, fish, and wildlife will be the foundation for people's
lawsuits against these agencies.

Chapter Twenty-one

The Century of the Farm and the Century of the Fish
William T. Davoren

This unusual meeting location for the Senate Committee on Agriculture and Water Resources may signify the arrival of the "Century of the Fish." This new century began about three years ago with the discovery that the subsurface agricultural waste drainage of the San Joaquin Valley did not consist of ordinary "salts," such as those found in the ocean or the bay, as we were told by the agricultural experts. We soon learned, nature's way, that these drain wastes include toxic elements such as selenium, boron, and molybdenum. These kill or maim animals or harm plants.

Thus did an unknowing society learn that giving agriculture the top rung on the water ladder could lead eventually to poisoning the well that nourished agriculture and California. Some historians trace agriculture's water dominance to the federal appeal court decision of 1884—the so-called Sawyer Decision—that outlawed hydraulic mining for gold as a public nuisance. The farmers took over from the gold miners at that point as the dominant force in California's politics and allocating water—though it took ten more years to really shut down "hydraulicking," as the most destructive form of making a living on California's frontiers was called.

So this began the "Century of the Farm," as far as water use in California is concerned. This century began to die in 1979, when

This statement was originally delivered to the California State Senate Committee on Agriculture and Water Resources meeting, Antioch, December 10, 1986.

the three "responsible agencies" decided that all of the San Joaquin Valley drainage, with the drainage of the Tulare Lake basin to come later, could and should be discharged into San Francisco Bay at Chipps Island. Completion of this two-hundred-mile drainage canal would have raised the curtain on the second Century of the Farm.

Fortunately for the bay, and for the Century of the Fish, inflation and a growing public suspicion of the diseconomies of the large federal and state water projects—plus the passing of the wave of federal and state officeholders who had built the infrastructure for completing the Century of the Farm between 1935 and 1979— made it possible for the Century of the Farm to begin to give way to the Century of the Fish. Progress has been slow but steady.

The political defeat of the Peripheral Canal in June 1982 strengthens the case for 1982 being classified as the last year of the Century of the Farm. Or the first year of the Century of the Fish. Some observers may prefer 1983 as the opening year of the Century of the Fish, as that was the year the creators of the Kesterson hallmark, the Bureau of Reclamation, began accepting the fact that selenium—in all its forms—is real and that drainage and dilution may not be an acceptable solution to the valley's problems of farm pollution. Kesterson's fishes were the first messengers; the birds came with their message in 1983; small mammals contributed their data in 1984.

Applying ancient methods of dealing with farm drainage problems—but on an exorbitant scale to match California's topography, social history, and political climate—will not work. Now we know that dumping subsurface agriculture drainage wastes anywhere in the Bay/Delta environment would be the final, ignorant blow by modern man to destroy one of nature's greatest gifts: the annually renewing, self-sustaining, community of life of the Bay/Delta estuary that brings spirit, joy, and economic enrichment to the lives of all Californians.

The Century of the Fish can restore the bay. At the same time it can restore California's confidence in its own public agencies. The recommendations that follow, if applied by this committee, can help clean up the problems left over from the Century of the Farm. California's new interest is to make sure that our three branches of government begin functioning sensibly to maintain the momentum of the Century of the Fish. We need ninety-six more years, starting now.

We recommend that this committee use its considerable influence to help protect and buffer the actions on the Bay/Delta problems that must be completed in the next four years by the State Water Resources Control Board. The preliminaries for these historic hearings are now behind us. The thirty-one-page workplan for the hearings has been issued. The board acting as best it can under the legislation that created it, in 1967 and 1969, and using the guidance of the State Supreme Court (*Audubon, Racanelli*) emerging in the past three years, will need all the protection it can get— from the legislature and from the administrative agencies—to produce the long-awaited decision on Bay/Delta water quantities and qualities. California has postponed, until this year, resolution of the Bay/Delta issue. In an ideal society the Bay/Delta issue would have been resolved *before* the large federal and state water projects were constructed in the past fifty-one years. That did not happen because our civic temperament doesn't tolerate delay. At this point the subsidies and economies of scale offered by the large public projects also offer the destruction of the remaining annually renewable values of our river systems, symbolized in this statement in the form of the Century of the Fish, as well as the destruction of water quality for all Californians.

Specifically, we request this committee to support any budget requests of the State Board for funds to make the Bay/Delta hearing process as open as possible to public participation. The fishery values already sacrificed in the Bay/Delta system amount to $2 billion since the federal and state project pumps began serious depletions and egg and larvae destruction in 1955. That estimate is for salmon, steelhead, and striped bass only. Annual losses amount to at least $117 million, according to Meyer Resources, Inc., in a report prepared for the Department of Fish and Game. Costs for involving the public in the hearing process are minuscule in comparison. The people of California, through the legislature and the polling place, have spoken out before to help guide this state's water decisions at critical times. All twenty-three or twenty-four million of us should be able to participate in the State Board's historic hearings, to the extent this is possible, in some way.

We request that this committee, and the legislature, apply their considerable influence on the administrative branch to encourage that branch to support fair, statesmanlike, and cooperative performance on the part of the major participating agencies involved in the Bay/Delta hearings. This applies primarily, of course, to the Department of Water Resources and the Department of Fish and Game. Both of these agencies, incidentally, are months or years behind in completing the research on Delta outflow/San Francisco Bay issues that the State Board ordered them, in 1978 (D 1485/ Delta Plan), to have ready for the hearings now getting under way.

Such bilateral agreements—or "trilateral," if you include the State Water Project Contractors Association—as the present one, approved by the two departments to resolve some poorly defined "mitigations" of SWP Delta pump damages to fisheries for $15 million, should be shelved for the duration of the State Board's Bay/Delta hearings. There are other trilateral agreements in this same class. There seems to be a steady stream of agreements being worked on in Sacramento by the DWR/DFG/Contractors trio. Now such bilateral or trilateral contracts should be tabled for four years. This is no time for business-as-usual, and it is no time for administrative agencies to be making deals to undercut, preempt, or circumscribe decisions the State Board must make by 1990.

Similarly, this Senate committee, or the legislature as a whole, may have to do something to help protect the State Board hearings from the machinations of the U.S. Bureau of Reclamation. Perhaps a resolution requesting the secretary of interior or the Congress to rein in the Bureau of Reclamation's regional director is advisable. That agency's use of the *approval* of the Coordinated Operation Agreement (in HR 3113) as a launching pad to sell another million acre-feet of stored water—under the guise of new "water marketing" strategies whose time has come—is both premature and premeditated. Obviously the bureau is attempting to *give* and to *make* commitments before any State Board decision can require the bureau to help carry the load of unmet water quality and fish and wildlife needs that might be assigned to it by the State Board decision in 1990. The bureau's unmet fish and wildlife obligations are documented. These failures stretch from the Trinity River to Fresno County and now include San Francisco Bay.

Chapter Twenty-two

The Central Valley Project and the Public Trust Doctrine
William D. Sweeney

My name is Bill Sweeney. I retired from the U.S. Fish and Wildlife Service a little less than two years ago, after thirty-three years. At the time of my retirement, I was the service's area manager for California—a position I had held since state-level area offices were established in 1976. Those offices were abolished in one of the first (and one of the worst) decisions affecting the Fish and Wildlife Service early in the Watt–Arnett regime at Interior.

I suspect you've already heard and read more than you really wanted to know about selenium, and I don't know much about it anyway—except, like sex, a little bit is good for you but too much could kill you. For you fatalists who figure, "What the hell, I've got to die of something," I recommend the former. All of my research convinces me that croaking from overindulgence with the opposite sex beats selenium poisoning by a mile. However, let's not declare the discovery of selenium in San Joaquin Valley drainwater an unmitigated disaster just yet. Perhaps its appearance, and the well-founded scare it has caused, will serve to focus attention on the bigger problem of how we are squandering our clean water resource and, with it, the fish and wildlife legacy which depends on that resource.

From a speech presented at a public meeting on "The Use of Agricultural Drainage Water on Private and Public Wetlands for Waterfowl" at Los Banos Fairgrounds, October 6, 1984; reprinted with permission; epilogue, August 1989.

My premise is this: Construction and operation of the Central Valley Project, spawned of a political alliance among the Bureau of Reclamation, the giant agribusiness industry centered in this valley, and certain present and former members of the U.S. Congress, has (1) resulted in unparalleled destruction of migratory bird and anadromous fish resources for which the federal government holds a trust responsibility and (2) turned the San Joaquin River into the lower colon of California—a stinking sewer contaminated with salts, heavy metals, trace elements, and the residue from the annual application of hundreds of tons of insecticides, herbicides, and fertilizers.

All of you duck hunters and fishermen are helping to subsidize this project. We pay in at least four different ways. First, the government sells water and power for a fraction of what it costs to develop. Then we pay again to support the prices of surplus crops produced; again in the loss of our fish and wildlife resources; and again for living in, and trying to clean up, a fouled environment.

Is it possible to effectively cope with the problems the Central Valley Project has created, and if so, how do we go about it? I think it can be done, but the process will be difficult, expensive, and time-consuming. One thing is certain—at least in my opinion—it will never get done if we wait for the state and federal agencies responsible for natural resource conservation to do it for us. They simply will not prevail against the water development agencies and their well-financed supporters in the agribusiness industry.

Fish and wildlife resources have suffered immense losses as a consequence of water development works. Although larger, more costly, more heavily subsidized, and more destructive than most, the CVP is only one among hundreds of similar federal boondoggles. In nearly thirty-five years of professional experience as a field biologist and as an administrator attempting to cope with these projects throughout the western United States, I can state without equivocation that fish and wildlife *always* come out on the short end of the stick. Despite tens of millions of dollars spent on biological field studies, tens of thousands of "coordination" meetings with water developers, thousands of reports, more thousands of recommendations, beaucoup pledges from politicians that the natural environment will be protected, fish and wildlife resources take a severe beating every time one of these projects is built. That's simply a stark political reality.

As a "for instance," twenty-five or thirty years ago the salmon run in the San Joaquin basin exceeded one hundred thousand fish. It's now down to five or ten thousand, mostly supported by hatcheries, with no natural spawning in the San Joaquin itself. That represents a loss to the offshore commercial troll fishery of at least two hundred thousand fish—equivalent to 1.5 to 2 million pounds worth $3–5 million at dockside. The commercial salmon fishing industry is going belly up, while we pay for more surplus cotton and feed grains. These crops are being grown by taking huge quantities of water swiped from our migratory birds and anadromous fish and applying it to land, much of which is alkaline desert that should never have been plowed. I tend to gag every time BuRec refers to the slopes on the west side of the valley as "prime" agricultural land.

Okay, you say. We know all that. We've heard it before. But if the state and federal environmental agencies are overwhelmed by the political clout of the water hustlers and the agribusiness industry, where do we turn for help? In my view, the time has come to turn to our last, and best, hope—the courts, particularly the federal courts.

Private environmental organizations such as the National Audubon Society, Defenders of Wildlife, National Wildlife Federation, Natural Resources Defense Council, and many others are today the country's first line of defense against polluters and exploiters. They're turning to state and federal courts with increasing frequency—and they're winning many more battles than they're losing. NRDC just took on Bethlehem Steel for polluting Chesapeake Bay in flagrant violation of its discharge permit—while EPA stood around and did nothing.

The Audubon Society, in its Mono Lake case against the Los Angeles Department of Water and Power (LADWP), which went all the way to the U.S. Supreme Court, established that the public trust doctrine is alive and well in California. This occurred despite the intervention of the U.S. Justice Department (at the request of BuRec) on the side of the LADWP. Mark this down in your little black notebook—the Mono Lake case is the most important water issue to be litigated in California in many years, and it was a private conservation organization, the National Audubon Society, that had the moxie to bring it to court.

As an aside, you may be interested in learning that as a result of testimony in early 1982 by me and members of my staff in support of the public trust doctrine and certain other water issues, Fish and Wildlife Service field personnel were forbidden by the service director to speak in any form on the subject of California water rights, state or federal legislation or initiatives affecting water rights, or the public trust doctrine. That prohibition, which is still in effect today, was generated within the department by the Bureau of Reclamation, which piously claims it would never attempt to muzzle a sister agency.

Unfortunately, there is no longer any focal point of accountability for the Fish and Wildlife Service in California. The responsibilities that were mine as area manager are now divided among a half-dozen or more people sitting safely out of the line of fire some six hundred miles away in the Portland, Oregon, Regional Office. And current political realities dictate that they listen much more carefully to the voices of the Interior Department's political administration three thousand miles away in Washington than to the cries for help coming from the environmental community in California.

The Fish and Wildlife Service's field people in California, and the Regional Office folks in Portland, are just as capable and dedicated as at any time in the past and there's plenty they'd like to say; but you're not going to hear it on the record. And while service personnel are forbidden to open their mouths on California water issues, no such constraint has been laid on Reclamation. Its officials are free to run up and down the Central Valley and say anything they want. So if you want your air and water and fish and wildlife resources protected, get a good lawyer and prepare to charge into federal court.

Okay, you say again. Lemme at 'em! But what laws are being trifled with, who do I sue, and what are the grounds?

Let's begin with the public trust doctrine. A large body of case law has been developed over the years based on application of this doctrine. For one of the most comprehensive discussions of the subject I'd recommend reading "The Public Trust Doctrine in Natural Resource Law: Effective Judicial Intervention," published in a 1970 issue of the *Michigan Law Review* by Professor Joseph Sax.

But for right here and now, one of the things that doctrine means to me is that there are public trust restrictions on the power of state to inordinately reduce streamflows. But that's not just my opinion. Ron Robie, in a paper delivered when he was director of the State Department of Water Resources, pointed out that the waters of the state are owned by the people of the state and that the state may not lawfully surrender title or control of that water in any way inconsistent with the administration of the trust under which that title is held. While the state can transfer the right to *use* the water, it cannot do so if it defeats the public rights in the trust.

I believe a defensible legal position can be taken that all water rights granted by the state are subordinate to the paramount duty of the state to carry out its public trust responsibilities. The Mono Lake case supports that view. If an existing water right results in the derogation of trust interests, it can be argued that sufficient water to protect those interests was *never* transferable. Migratory birds and anadromous fish are trust resources. Shutting off the natural flow of the San Joaquin River through construction and operation of Friant Dam and turning that stream into a sewer pipe for irrigation drainwater has derogated the very hell out of those resources. I believe the Bureau of Reclamation and the State Water Resources Control Board are vulnerable to a legal action brought under the public trust doctrine. If someone tackles them on this issue, I think there's a good chance to get some clean water back in the river.

There are other avenues that present real possibilities for legal challenges. Migratory bird treaties with Great Britain, Mexico, Japan, and the USSR have been signed by the president and ratified by the U.S. Senate. The Congress has enacted the Migratory Bird Treaty Act and other legislation implementing these treaties— which provide for the conservation and development of migratory birds and their habitats—in recognition of the trust responsibility of the federal government.

Construction and operation of the Central Valley Project has resulted, both directly and indirectly, in enormous losses of migratory bird habitat. Those massive reductions in habitat are directly reflected in fewer birds. In addition, the poison in the San Luis Drain is killing birds directly, God knows how many. I believe this constitutes a "taking" of migratory birds—a violation of the Migra-

tory Bird Treaty Act, just as surely as that of a poacher sneaking out
and ground-sluicing a few sprig (shooting sitting ducks) thirty minutes before shooting time.

Is a water right granted to Reclamation inviolate even when
exercise of that right is adverse to the health and welfare of migratory bird resources for which the federal government has a trust
responsibility? Is a treaty signed by the president and ratified by
the Senate on behalf of migratory birds somehow inferior to a
contract signed by Reclamation to deliver water to a private user?
I'd love to see these questions argued in court. Oil companies and
other corporate polluters have been successfully prosecuted under
the Migratory Bird Treaty Act. Why not BuRec or the Westlands
Water District?

Then there's the Endangered Species Act. Just about every one
of the several San Joaquin Valley critters on both the state and
federal lists are endangered because massive expansion of irrigated
agriculture has nearly wiped out their habitats. The law not only
prohibits federal agency actions that are inimical to endangered
species but also mandates those agencies to modify their programs,
whenever possible, to benefit endangered species. I've seen little
evidence that Reclamation is very much aware of its responsibilities
under the Endangered Species Act, much less in actual compliance
with it. Maybe a suit brought under Section 9 of that act—the
"taking" provision—would get their attention.

At the state level, the Water Resources Control Board has not
developed water quality standards for irrigation drainwater. Why
not? No municipality or industry can dump the kind of stuff into the
state's waterways that frequently is discharged or leaks from many
irrigation drainage systems. I'm not saying that no city or industry
ever dumps bad stuff. We all know they do it now and then and get
away with it, but it's done in violation of their discharge permits
and it's prosecutable. For irrigation drainwater, there aren't even
any standards, much less a permit system that can be enforced in
the interest of public health and safety, to say nothing of the public's fish and wildlife resources. Somebody ought to be asking the
State Water Resources Control Board to do its job.

The California constitution requires that waste or unreasonable
use or unreasonable method of use of water be prevented. I believe
a strong argument could be made in a court of law that it is *unrea-*

sonable to apply vast quantities of water to certain lands—not to provide the moisture needed for the crop but solely to flush salts and other poisonous chemicals from alkaline soils closely underlaid with impervious layers of clay and then to dump that dirty water on people downslope and downstream. It would be interesting to see that issue argued in state courts.

Right down here at the local level, how many of you members of Grasslands Water District hunting clubs have become thoroughly familiar with the delivery and drainage systems that control your water and appreciate the fact that managing wetland habitat with irrigation drainwater is a complex and risky business?

Six years ago, when I was area manager, service personnel here in California recognized the need for detailed research on the use of drainwater in San Joaquin Valley marsh management. No waterfowl biologist in his right mind would accept, without reservation, the extravagant claims being made by proponents of a valley drain regarding the benefits to waterfowl of all the dirty water soon to be available.

With funds scrounged from ongoing programs, we put together a plan of study. As long ago as 1980, the Portland Region of the FWS highlighted drainwater research as its highest priority. We couldn't get a dime from BuRec, the SWRCB, DWR, or Fish and Game to help fund it. What was even worse, we couldn't even get our own Washington office to put anything in the budget for it. Now that the manure, or the selenium, is in the fan with respect to the long-term effects of drainwater on waterfowl habitat, everyone wants to get into the act. A "bobtailed" version of our original plan of study is now being rushed into action with funds to be made available by many of the agencies that couldn't help three or four years ago. Unfortunately, the three years of water quality monitoring called for in the original plan, badly needed to provide baseline data as a point of departure for experimental marsh management techniques, was never carried out. Now the service is leaping into a crash program without the time, the funds, the specially trained personnel, or the background data to do it right. It will probably yield the results common to most crash research efforts—a conclusion that more study is needed.

The Fish and Wildlife Coordination Act, even though it hasn't worked worth a damn at protecting fish and wildlife resources from the ravages of federal water projects, does offer some possibilities for judicial review. The act imposes some very clear requirements on federal agencies—both the Fish and Wildlife Service and the construction agencies. It mandates "equal consideration" for fish and wildlife resources in project planning, close coordination between construction and fish and wildlife agencies, reports on measures to compensate for fish and wildlife losses, and specific recommendations to Congress.

Only infrequently are all of these steps followed in a timely fashion, and when they are, Congress more often than not simply looks the other way. But according to Oliver Houck, former general counsel to the National Wildlife Federation, in an article in the *Environmental Law Reporter,* there are a number of avenues through which to seek judicial relief under this statute that have not yet been carefully and thoroughly explored by the legal profession. The Central Valley Project, in my view, presents some marvelous opportunities to test Houck's suggested approaches.

Recommending that you haul the bad guys up in front of a judge probably strikes you as a hopelessly impractical solution to your water problems. I'm well aware that not many individuals have the money, the time, or the persistence to take on a government agency or one of its contractors in court. The big advantage the agencies have is literally unlimited money and time to defend themselves. They can turn hordes of staffers loose developing data and testimony to prove that black is white. And there are no meaningful constraints on how much staff time and taxpayers' money can be spent by the government, state or federal, in its own defense.

But perhaps a few of you can afford it and do have the time and patience to see it through. I understand one person with certain interests down here has started legal proceedings. I'm not familiar with the specifics of his allegations, but I hope he's zeroed in on the right agencies for the right reasons. Perhaps one or two more of you feel strongly enough about what's going on to institute a court challenge.

For those of you who can't afford to go it alone, join one of the organizations that do litigate environmental issues and urge that organization, on its own or in concert with others, to come to your aid. The state and federal water development agencies and their big water contractors are zealous in promoting and defending their narrow objectives. Despite much rhetoric, they're not nearly so zealous in cleaning up the environmental messes their projects leave behind.

It's taken nearly twenty years to push Reclamation into doing something about the monstrous salmon and steelhead losses resulting from the Trinity River Division of the CVP. Many of the problems on the Trinity were predicted by fish and wildlife biologists before the project was built. The diversion dam at Red Bluff on the Sacramento River, in operation since 1966, has been a known anadromous fish bottleneck for over fifteen years; but only during the last year has Reclamation started to do something about it. Nothing much has actually happened yet on either the Trinity or at Red Bluff; but for the first time in nearly two decades it appears that something might. In the case of the Trinity, special legislation is slowly working its way through Congress to force Reclamation to clean up its mess. The legislation, wouldn't you know it, only requires the water users to pay half of the cost. The rest of us taxpayers have to pay the other half—to ransom back our own fish and wildlife resources. The important lesson here is that in both cases, the Trinity River and Red Bluff Diversion Dam, private individuals and organizations outside of state and federal government have been instrumental in forcing these issues.

When it comes to cleanup, the water agencies will do only as much as they're forced to do, and the best forcing mechanism is an outraged, and organized, private citizenry. There's a water-related crisis in the San Joaquin Valley, and selenium in the San Luis Drain is just one tentacle of the octopus. Chop it off and a new one will grow. Huge groundwater overdrafts. Taxpayer-subsidized water, taken from our migratory birds and anadromous fish without even token compensation, applied to alkaline soils to grow surplus crops. Irrigation drainwater, all of dubious quality and some of it downright poisonous, coming off the land in increasing volume every year. The San Joaquin River—which once supported tens of thousands of spawning salmon and whose overflow waters nourished thousands of acres of wetlands—now a public disgrace.

Lock arms and say, "I'm damn mad, and I'm not going to take it any more!" You can't turn this valley back to what it was a hundred and fifty years ago, but you can make it a hell of a lot better for fish and wildlife than it is now. You've got some public trust rights as citizens that have been trampled upon. And you've got the legal tools to obtain a fair measure of redress. How it all comes out is up to you.

Epilogue

Nothing has happened since I delivered my paper in October 1984 on the "Central Valley Project and the Public Trust Doctrine" that would alter its original thrust:

1. Reclamation has conned the State Water Resources Control Board into permitting a minimal effort at cleaning up Kesterson Reservoir.

2. The State Board, prompted by screams from Reclamation, the San Joaquin Valley agribusiness lobby, and southern California's Metropolitan Water District, rejected its own staff's report on how to make things a little better in the Delta.

3. Reclamation has attempted (unsuccessfully for the moment) to ram through three half-baked environmental impact statements supporting its proposal to sell the last remaining drop of uncontracted CVP water (if, in fact, there really is a drop left).

4. Nearly ten thousand acres of evaporation ponds in the Tulare basin, as contaminated as Kesterson Reservoir, are killing and deforming migratory birds daily—in direct violation of the Migratory Bird Treaty Act.

5. A new secretary of interior, Manuel Lujan, says he can do nothing to prevent the renewal of Friant Unit water contracts—the original issuance of which, forty years ago, resulted in the total destruction of the San Joaquin River and its run of nearly one hundred thousand salmon.

6. That same Lujan, with convoluted and tortured reasoning, approves a smelly proposition put forth by the J. G. Boswell Company that will permit Boswell (like many other big San Joaquin Valley landowners) to continue to violate certain conditions of current Reclamation law.

7. The regional director of the Fish and Wildlife Service in Portland is forced out of office for having the temerity to take a few modest stands in favor of fish and wildlife resources.

The list goes on and on. Most of these issues will only be finally resolved in federal court. At least one of them, the fifth item, is already being litigated. The Natural Resources Defense Council and the several other environmental organizations that have joined with the council will win that suit. Of that I have no doubt. The Bureau of Reclamation, the State Water Resources Control Board, the L.A. Department of Water and Power, the Metropolitan Water District, and all of the big landowners and water hustlers up and down the Sacramento and San Joaquin valleys can be beaten in court, because they have violated the people's rights to its water. Let us hope that the funds and the resolve of the private environmental community hold out long enough to get the job done.

Restoration Efforts
Local Perspectives

Part Four draws attention to a variety of approaches to fishery restoration practiced in California today. They range in tone from Richard J. Hallock's fact-filled, dispassionate discussion of Sacramento River problems and opportunities to Ken Hashagen's piece on the Salmon Stamp Program and locally oriented efforts as exemplified in the concluding chapters.

It is intriguing to consider these approaches in terms of people involvement. Restoration efforts in the Central Valley, the major salmonid-producing area in California, seem to rely on government-employed professionals and commercial fishermen. Coastal streams evoke a much more intense, personal involvement in restoration work. Why the difference? Two reasons appear likely. First, the Central Valley is vast and decentralized, its communities agricultural; in contrast, coastal basins and their smaller, diverse communities are much more cohesive. Second, large artificial-reproduction facilities associated with the Central Valley Project and the State Water Project tend to promote reliance on government and quasi-governmental agency solutions to fishery problems; in contrast, small, localized projects elicit more personal involvement. The significance of these differences—and whether anything should, or could, be done about them—is an open question. Some fishery professionals are convinced that a more personal approach would help Sacramento River stocks.

Chapter 23 is the abstract of a comprehensive report by Richard J.

Hallock, retired associate CDFG fishery biologist. When the Advisory Committee conducted its survey of stream resources, Hallock agreed to prepare a study of Sacramento River conditions. His report stands out as an excellent example of the results of that survey. The abstract of Hallock's report reproduced here reveals the problems facing salmonid resources and predicts possible beneficial effects if these problems are resolved. He suggests that professional "stream managers" on several smaller Sacramento River tributaries would increase production.

Would you ask to be taxed? Ken Hashagen, in Chapter 24, tells about salmon fishermen who did just that. His subject is the Salmon Stamp Program, a unique approach to fishery restoration initiated by commercial fishermen. Not so many years ago, anyone with ten dollars could buy a commercial fishing license. That has changed. Now new entries into commercial salmon fisheries are restricted in number, and the cost of a license to fish solely for salmon may reach $505 each year, much of which pays for salmon restoration projects selected by the fishermen themselves. Commercial salmon trollers initiated this program in the 1970s, when the Department of Fish and Game agreed to match trollers' contributions. Although the state contribution is far less now, the program continues to raise in excess of a million dollars a year.

Chapter 25 reintroduces Scott Downie (Chapter 3), who discusses several current restoration projects involving fishermen, most notably the Cooperative Resource Management Program (CRMP) that evolved from fishermen's initiatives. This holistic program looks at salmon restoration as part of a broad spectrum of basin restoration needs involving public and private lands, the timber industry, community workers, and government agencies. In his conclusion, Downie revisits the East Branch "sixty years and one mile" from where Edith Thomas challenged neighbor Elmer Hurlbutt to a fishing contest. His description of present conditions in that watershed establishes beyond question the pressing need for basinwide cooperative approaches to restoration.

The California Conservation Corps is the subject of Chapter 26. Here the focus is the Humboldt Fire Center near Weott, where a number of crews have been involved in the Salmon Restoration Project, led by the Department of Fish and Game. By working cooperatively with industry, landowners, various government agencies, and fish restorationists, the CCC is achieving notable results, marked most dramatically by reports that fish are now spawning in North Coast streams virtually barren for many decades.

The three concluding chapters describe specific coastal restoration activities and examine the potential for urban stream renewal in the San Francisco Bay region. The scene of Chapter 27 is California's "lost coast," a desolate, wind- and rain-swept area clinging to the southwestern edge of Humboldt County, a place where people go to recharge psychic batteries depleted by urban life. Here the Mattole Restoration Council coordinates attempts to rebuild salmon runs in a watershed nearly destroyed by tan-oak harvesters, ranchers, and lumbermen. Low-cost, low-tech methods are employed, and the ubiquitous fifty-five-gallon drum and miles of plastic pipe are utilized at various sites. Streams are made suitable for spawning. Slipping land is stabilized by planting native greasewood. Cooperation among the Department of Fish and Game, county government, local landowners, and land users is evolving, hopefully to the full-fledged CRMP level that Downie describes in Chapter 25.

In Chapter 28, the scene changes sharply. The restorationists William Hooper tells about in his essay have a different dream: they would restore the streams in their city backyards. In this endeavor they work closely with other urban restorationists. The focus generally is not on the fish themselves but on the re-creation of natural beauty where ugliness long ago became the norm. While fish populations are perhaps only 5 percent of what they once were, salmonids are still found in many Bay Area streams, such as Petaluma River, north of the bay, and San Francisquito Creek, which flows through Stanford University campus. Because fish populations are so small, restoration efforts are commonly private, although some funding from the state Department of Water Resources is available. DFG personnel provide helpful advice. But urban fishery restoration is not confined solely to restoring beautiful fish in enchanting settings. Urban fish restorers seek opportunities for classroom education and enlightenment of the general public regarding statewide fishery conditions.

In Chapter 29, "Saving the Steelhead," Eric Hoffman captures another element in salmonid restoration activities. On Big Creek, near Davenport, landowners with several generations of tenure have taken the initiative. Their memories of historic runs, and poaching with long gaffs while keeping a lookout for Grandfather, who was sworn to enforce the law, constitute an enduring legacy that must not be lost. The Big Creek project, like others up and down the coast, demonstrates again these fishes' peculiar capacity to make humans feel protective, a quality remarked upon by other contributors to this volume. Like those other projects, too, this one shows that the intense interest of one or two people, coupled with professional expertise, is capable of galvanizing broad community support. People do make a difference.

Sacramento River Problems and Opportunities

Richard J. Hallock

Total numbers of salmon that spawn in the Sacramento River system have declined more than 50 percent (75 percent in the upper river) since the 1950s. Fall-run salmon, which make up more than 90 percent of the total, appear to be stabilized at a low level of two hundred thousand fish; 85 percent spawn naturally and 15 percent are spawned artificially at hatcheries. On streams where there are hatcheries, however, the populations are increasing, which is masking the true picture; that is, the natural spawning populations are declining in the upper Sacramento River system (above the Feather River).

Most of the known problems in the Sacramento River system, now limiting salmon and steelhead production, occur in the upper Sacramento River and are apparently adversely affecting the natural stocks much more than the hatchery stocks. The two most important known recent causes of the salmon declines in the upper Sacramento are Red Bluff Diversion Dam (RBDD), which has reduced salmon populations by 114,000 fish, and the Glenn-Colusa Irrigation District (GCID) diversion, which has reduced the salmon populations by 35,000 fish. Between the two, they could be depriving the fisheries of 300,000 salmon at a two-to-one catch-to-escapement ratio.

A combination of mining pollution, flow fluctuations and warmwater releases from the Shasta/Keswick dam complex, lack of suitable spawning gravel and gravel recruitment, unscreened diversions

These nests (redds) of spawning chinook salmon in the Sacramento River were exposed when water releases from Shasta Dam were reduced. Dewatering of redds elevates water temperatures to levels that are lethal for egg survival. A single redd dug by a female salmon may cover forty square feet of streambed and contain more than six thousand eggs. (California Department of Fish and Game)

(as well as inefficiently screened diversions), predation, and operation of the Anderson-Cottonwood Irrigation District (ACID) diversion dam and RBDD causes an estimated 85 percent loss in natural stocks between eggs deposited in the gravel and smolts entering the ocean. The loss is not as great for hatchery production, partly because the size of fish released is greater and a large portion of the production is released at downstream sites or in San Francisco Bay.

Restoration of salmon populations in Clear, Battle, and Butte creeks could increase the salmon populations by 17,500 and steelhead by 1,300 fish. This salmon restoration could increase the fishery landings by 35,000 fish at a two-to-one catch-to-escapement ratio. There are seventeen smaller Sacramento River system tributaries that now support a combined population of 9,000 salmon and 2,500 steelhead and are contributing 18,000 salmon to the fisheries. The problems are many, but one way to help assure continued or increased production on these streams would be to assign a stream manager to oversee the populations from the time the adults enter the streams until the juveniles migrate out.

Carrying out the proposed plans to expand artificial production at four Sacramento River system facilities could increase total hatchery production by 70 percent, from the present 44 million to 74 million smolts, subyearlings, and yearlings. There would also be an increase of at least 300,000 in yearling steelhead production. Based on the current spawning population size of 200,000 fall-run salmon, the natural spawners would still be producing about 70 percent of the juvenile outmigrants and the hatcheries 30 percent, but the hatchery fish would be much larger. Before going beyond this point with increasing hatchery production, it would be appropriate to examine the effects of this increased production on the natural stocks.

The greatest potential dangers to salmon and steelhead production in the Sacramento River system include the anticipated water conditions in the year 2020 (when final Bureau of Reclamation water contracts will have been let), proposed power projects at ACID and RBDD, and continued bank stabilization with rock riprap.

Chapter Twenty-four

The Salmon Stamp Program
Ken Hashagen

California's commercial salmon fishermen make a living off fish that belong to the people of California—your fish and my fish. It's a tough, hard job with long hours under sometimes terrible weather conditions, and the pay isn't great, but those are *our* fish. Sure, they have to buy a vessel permit and a license, but then we have to have a fishing license to fish for trout or bass, but we can't sell our catch to make a living. How can the Department of Fish and Game let this happen?

Actually, California's commercial salmon fishermen are paying their fair share and more for their use of a public resource. Historically, ocean troll fishermen began fishing for salmon about the first of May and fished until the end of September. They caught and sold as many salmon as they could during the season. There were relatively few boats, and fishing gear was primitive by today's standards.

Then salmon populations began to decline—dramatically—from poor logging practices, which put debris and sediment into the steams; from dam building, which cut fish off from historical spawning areas; from gravel mining, which removed spawning gravels; from road building, which increased sedimentation and damaged spawning gravels; and from more people fishing and using more sophisticated boats and gear. Because of this decline, the California Department of Fish and Game began a program in the late 1970s to

This essay originally appeared in *Outdoor California*, a publication of the California Department of Fish and Game, September–October 1987; updated June 1989.

reduce the size of the salmon fishing fleet; it also shortened the length of the fishing season in an attempt to reduce the commercial and sport catch, which would allow more adult salmon to return to the rivers to spawn.

When your livelihood is threatened, it's time to take action, and the commercial salmon fishermen did. As a group, they requested the legislature to pass legislation that would, in essence, tax them. Assemblyman Barry Keene authored AB 2956 in 1978; his bill required each person fishing commercially for salmon to purchase a salmon stamp costing as much as $30 each year. The money collected was used to rear one million yearling chinook salmon. The legislation required the Department of Fish and Game to match the trollers' contribution by rearing and releasing an additional one million yearlings each year. This innovative piece of legislation has accounted for at least sixteen million additional salmon released into the ocean for the use of both commercial and sportfishermen and, as well, return to the river for spawning.

Pleased with this program and not content to rest on their laurels, the trollers requested additional legislation in 1982. Mr. Keene introduced SB 782, which increased the cost of the initial stamp by $55. This augmented stamp had a provision that raised the cost of the stamp by $10 for every 250,000 pounds of salmon landed the previous season in excess of six million pounds. The cost could not exceed $215 per year. The funds collected were placed in a special account in the Fish and Game Preservation Fund and could only be spent for new or expanded salmon restoration and enhancement projects. A committee of four commercial salmon fishermen and one representative from the department met periodically to develop an annual program to expend these funds.

The program was reviewed and approved by the director of the department and included in the department's annual budget. Passed in April 1982, this legislation was scheduled to end January 1, 1987. In this five-year period, the dollars generated (approximately $2.7 million) funded many worthwhile projects, including the following.

Expansion of yearling chinook facilities at Mokelumne River fish facility. This two-year project included construction of an additional yearling pond, cementing of two ponds, construction of a "bird cage" to exclude herons and gulls, and a cover for the water aerator to control algae. A new feed truck was purchased and operating costs were paid annually. This facility can now raise two to three million more subyearling chinook salmon than before the renovation.

Trinity River yearling chinook program.   A total of $50,000 was used to rear five hundred thousand yearling chinook at Trinity Hatchery for the Klamath/Trinity river system in 1983–1984.

Construction of a trap for squawfish at Red Bluff Diversion Dam.   Since the construction of the Red Bluff Diversion Dam in 1966, large numbers of squawfish congregate below the dam and prey on downstream-migrating salmon reared in the river or released from Coleman National Fish Hatchery. The trap is used to reduce numbers of squawfish during their spring spawning migration past the dam. Juvenile steelhead also benefit from the reduction in squawfish numbers.

Construction of a permanent fish ladder and trap at Iron Gate hatchery.   This new ladder and trap were first used in the fall of 1984 and worked beautifully, greatly increasing the ability of the hatchery to trap and hold adult chinook and coho. Before construction of the new ladder and trap, salmon had the choice of entering a short wooden ladder and trap at the hatchery outfall or passing the hatchery and attempting to find the poorly designed entrance to a trap at the dam. The wooden trap was operated only during daylight hours because of safety considerations; the new one operates twenty-four hours a day. Hatchery personnel estimate that well over fifty percent of the 1984 run which reached the hatchery area entered via the new ladder, a far higher percentage than ever used the other two traps.

Technical assistance for fish culture projects.   A full-time fish culturist is paid for by Salmon Stamp funds to provide technical assistance to all public mini-hatcheries and pond rearing programs in California. He makes recommendations on the operation of the various facilities and disease treatment and organizes a spawning workshop each fall for new personnel.

Operation of Thermalito afterbay ponds.   These ponds, built as an annex to Feather River Hatchery, have the capacity to rear an additional six hundred thousand chinook yearlings annually. The ponds (actually raceways) were built as a cooperative DFG/DWR project. The Salmon Stamp Fund annually pays for five-sixths of the operating costs necessary to rear the additional fish.

Klamath/Trinity River hatchbox and pond rearing project.   Two sites (Horse Linto Creek and South Fork Salmon River) have been developed to trap returning adult salmon, spawn them, hatch the eggs, and rear the resultant fry to release. This project, funded the last four years, has excellent potential; however, low chinook runs have limited the project's opportunities to trap fish and operate the facilities at capacity.

Construction of Hollow Tree Creek weir.   Hollow Tree Creek is a tributary of the South Fork Eel River. The weir has been "temporary" for many years. Funds were used to build concrete structures for wooden dam boards in the new weir/dam, a fish ladder and trap, and a basket system to move adult fish to a holding area. Fiberglass tubs will be used to hold both adult fish and the resultant fry. A small one-room "house" was built for the personnel manning the trap and caring for the fry. Projected capacity is four hundred thousand chinook eggs annually.

Upper Klamath River habitat restoration projects.   The Salmon Stamp Program funded a number of projects in the upper Klamath River drainage, including placement of spawning gravel at Bogus and Beaver creeks, removing a rock barrier on the South Fork Salmon River, construction of five small fishways around diversion dams in the upper Shasta River, removal of a one-fourth-mile-long logjam in Mill Creek, construction of screens for two unscreened diversions, and rechannelization of lower Shackleford Creek to concentrate and deepen flows.

Purchase of miscellaneous hatchery equipment.   These purchases include a fish transport tank/trailer, incubators, refrigeration units, and filters.

Production of salmon restoration video.   The Salmon Stamp Program funded production of two videotapes. One is directed at the commercial salmon fishermen to show them what their stamp funds have paid for and what the benefits of this program have been. The other video shows a wide variety of salmon restoration projects in California (not just those funded by the trollers) and will be directed throughout the West on public television to educate the public on the need for restoration and about the programs under way.

These projects and the funds generated for them indicate a substantial commitment by the salmon trollers to their industry. There is still more, however. In September 1986, the trollers renewed the legislation for the stamp program until January 1992, but they lowered the base amount from six million pounds and increased the fees to $12.50 for every two hundred and fifty pounds above the base (from $10). The end result will be even more money available each year for restoration programs. For example, landings last year were over 7.75 million pounds; therefore each commercial salmon fisherman will have to pay $260 to go fishing in 1987 ($30 for the initial stamp and $230 for the augmented stamp). Additionally, and significantly, commercial passenger fishing boat operators, who take sportfishermen out on the ocean to fish for salmon, have asked to join the program; they were part of the legislation (SB 2517— Keene) that became law in September 1986.

With the expanded and extended legislation, the Salmon Stamp Committee has continued its active program, providing funding for the following projects:

- Continued funding for the hatchbox/pond rearing programs on Horse Linto Creek and the Eel, Mattole, and Little rivers.
- Continued funding of operational costs at the Department of Fish and Game's Mokelumne River fish facility and Thermalito afterbay ponds.
- Provided funds (approximately $125,000) for purchase of equipment at several department hatcheries, including two 2,800-gallon transport tanks, a fish crowder, bulk feed bins, incubators, tanks, troughs, and a fish pump.
- Continued funding the fish culturist who provides technical assistance to mini-hatcheries and pond rearing programs throughout California.

- Began funding educational projects ($10,000 to $15,000 each year) that provide classroom aquaria, field trips, and teacher curriculum packages for many North Coast elementary schools.
- Funded the construction of a bird exclosure at Coleman National Fish Hatchery near Redding, where birds were eating a significant portion of the salmon production. The committee provided $317,000 to cover half the cost of constructing a permanent weir at Mad River Salmon and Steelhead Hatchery. The weir will direct returning salmon into the hatchery and enable the hatchery staff to take more eggs, rear more fish, and augment the chinook run in the Mad River. Construction is slated for the summer of 1989.
- Provided funding to the Department of Fish and Game to construct a trap for returning chinook salmon near Los Banos on the San Joaquin River. Low fall flows prohibit salmon from migrating out of the San Joaquin into the Merced River and into the department's hatchery. The trap (on Mud Slough) was installed in 1988 and worked well, providing many adults for the Merced facility. Improvements should make it even more effective in coming years.

Chapter Twenty-five

North Coast Salmon and Steelhead and Their Habitat (2)

Scott Downie

During the early 1970s northern Californians became aware that their salmon and steelhead stocks were in jeopardy. At this time the legislature's first Citizen's Advisory Committee on Salmon and Steelhead produced three reports that publicly and graphically documented the depressed status of the fishery. Many policy recommendations and legislative actions resulted from those investigations and reports.

On the North Coast they spurred a fledgling interest in trying to do something tangible to revive, or at least stabilize, the dwindling salmonid populations. In Fort Bragg, the Salmon Restoration Association, made up of salmon trollers, sportfishermen, timber interests, and local community members, instituted a program to deal directly with local restoration problems that were underfunded by the legislature and the Department of Fish and Game. They founded the annual "World's Largest Salmon Barbecue" to generate cash to augment volunteer and in-kind restoration efforts. The association and its barbecue continue to this day.

The program has grown from relatively small-scale fry rearing and egg hatchbox ventures on selected coastal streams to the current operation that includes the Hollow Tree Creek egg-taking station on that South Fork Eel River tributary. It now produces up to two hundred thousand smolts annually for the Eel River system. Early citizen efforts also emerged in Ukiah and Eureka and along the Mattole River, among others, at about this time. All were responding to what was perceived as an overwhelming and critical situation for both depressed fish stocks and a Fish and Game Department short of manpower and funds.

By 1982 a new alliance had developed to try to save North Coast fisheries. With guidance and supervision from the department, private fishery contractors became actively engaged in salmon and steelhead restoration projects throughout northern California. The legislature designated specific funds to support these largely volunteer programs. The commercial salmon trollers self-imposed a landing tax (the Salmon Stamp Program) that has generated up to $1.25 million annually to operate or augment salmon enhancement programs.

In 1983 at Bodega Bay a conference of interested citizens, representatives of public agencies, and salmon restoration workers led to formation of the California Salmon, Steelhead, and Trout Restoration Federation. The federation has since become a major source of energy and expansion for regional fishery improvement programs. Projects now exist in nearly every watershed in northern California. This growth has occurred largely due to cooperation and training from a Department of Fish and Game that increasingly recognizes the power and potential of a citizenry involved actively in constructive watershed and land use management. For example, in watersheds checkered with both public and private ownership, and therefore often varying management priorities, cooperative resource management programs are needed to achieve wise watershed management. These CRMP programs are now becoming more common and can deal with the cumulative effects of overall watershed land use.

The various projects, whether conducted by private groups, landowners, or public agencies, are still hard pressed to meet the decline experienced by the fisheries. Their efforts usually take form in three general techniques: artificial propagation, habitat restoration, and educational programs. The first, hatchery production, is commonly viewed as a panacea or quick fix. Results are often dramatic and highly visible: large numbers of fish are released into streams where natural populations long ago disappeared. We have learned, however, that unless the underlying factors responsible for the absence of stocks in a watershed are also dealt with, we are merely engaging in an artificial "put-and-take" fishery program that cannot endure over time. On the North Coast, coordinated habitat improvement programs can usually be designed in conjunction with bioenhancement schemes, ideally using surviving natural stocks as a broodstock. This approach ensures that the increased populations will have a good chance to establish themselves as a self-perpetuating run. Some current projects have demonstrated the validity of this approach.

These coordinated habitat programs are conducted in many forms. For example, many streams are devoid of habitat diversity: two few pools, "cemented" spawning riffles, and a general dearth of streamside vegetation and large instream structural elements. These instream logs and boulders are particularly important to provide shade and cool water, create pools, and supply protection from predators. Typical techniques for treating these habitat problems include riparian planting, streambank stabilization, and direct placement of boulders or log weirs. In addition to the many completed projects, an important outcome of these activities has been to increase the ability of citizen restoration teams to diagnose and treat habitat problems throughout watersheds.

Upper slope erosion and drainage control problems are also critical to the streams and their fish populations. These areas often require the cooperation of many basin landowners and land users. Here again revegetation is a powerful treatment. But usually on the North Coast, road systems with their associated drainage systems serve to intensify, accelerate, and concentrate runoff during freshets. Therefore, they often must be modified to achieve meaningful soil conservation. A related and growing problem, as human population increases, is domestic and agricultural water consumption. More and more straws are constantly being thrust into regional watersheds that seem to have ever dwindling storage capabilities. Education in wiser land and water usage is required to address these concerns.

Public and formal environmental education is increasingly focusing on our local North Coast salmon and steelhead situation. These fish are the sensitive barometers of a watershed's health and demonstrate to us major potential problems that can affect people as well as fish and wildlife. In some areas, for example, creeks that formerly had strong year-round flows and fish in them have been

Small structure, big effect. Headwater stream restoration by artificial placement of stone or log deflectors adds oxygen to the water and provides protective rearing habitat for juvenile salmonids. Thousands of such structures are being built on California river tributaries. (Andy Kier)

reduced to low and intermittent streams during the past twenty-five years because of poor basin conditions. For many years they have been unable to support summering coho salmon or juvenile steelhead.

Despite shrinking water resources, signaled by vanishing fish stocks, basins have been subdivided, developed, and populated without concern for fish and wildlife needs. Increasingly people are being affected: hauling water to homesteaders in these tapped-out basins is now a flourishing summer business.

Although the technology, and to a large degree the will, now exist to reverse the demise of the North Coast's salmonids, the task has only begun and the opposing forces are very active. Land subdivision and poor road construction abound. Competition for water increases daily in the name of progress. The day of conifer forests composed of large second-growth or old-growth trees, vital to the stability of the steep mountainous terrain of northern Califor-

Gullying is unmitigated waste. Loss of highly erodible soil in the Trinity River watershed destroys usable land and chokes salmon spawning habitat with blankets of silt. (Bureau of Reclamation)

nia, is probably gone forever: a sacrifice to short-term, intensive forestry practices intended to maximize immediate profit. Although modern timber harvest practices incorporate vastly improved and far less destructive technology than in the "bad old days," the timber industry in general demonstrates little interest in repairing the fisheries devastated by its first cut. That first cut, made years ago, is still physically harming our fish and streams, along with its legacy of marginal timber production at the expense of natural and healthy fish production.

Given a chance, these fish have proved to be hardy and "renewable" natural resources on a much faster turnaround than timber. Ironically, timber and fish production are both dependent on healthy environmental conditions that are complementary rather than at odds: good water and good soil, both in their correct and natural places, ensure good survival and production of trees *and* fish. Neither can exist in the long term by relying on hatchery clones or nursery hybrids to compensate for loss of soil, water, or crucial vegetative cover.

On one affected North Coast stream, the East Branch of the South Fork Eel, Edith Thomas's grandson operates a downstream migrant trap to compare its relative salmon and steelhead populations with similar streams in southern Humboldt County. From mid-April to mid-May 1988 he caught a *total* of only thirteen chinook fry, one coho fry, and nine newly emerged steelhead. (On control streams, *overnight* numbers trapped at that time were typically in the hundreds or thousands.) The East Branch, with twenty-seven miles of streambed, is the largest tributary of the South Fork. It is also one of the most severely impacted: by late May water temperatures are lethal to salmonids, and by July the stream is essentially dewatered.

These are the conditions that exist now—sixty years and one mile from where Edith Thomas exulted in her triumph over Elmer Hurlbutt after she had successfully enticed an eighteen-inch midsummer rainbow to accept her hand-tied grub larva fly. When you understand that, you will begin to comprehend the magnitude of what has been lost on the North Coast.

Chapter Twenty-six

The CCC's Salmon Restoration Project

Kim Price

California's salmon and steelhead benefit from a state program involving young men and women: the CCC. During the 1930s Depression, "CCC" meant Civilian Conservation Corps, part of a national recovery program intended to provide outdoor jobs for youth. Since 1976, California has had its own CCC: the California Conservation Corps, with a similar focus on developing the work ethic but with purposes that go far beyond merely getting youth "off the streets."

Throughout the state, some two thousand young men and women from eighteen to twenty-three years old are involved in a variety of jobs that take them to the most remote parts of rural California. In recent years, a significant segment of this group has been involved in the North Coast's Salmon Restoration Project, a joint endeavor with the Department of Fish and Game and local organizations. Along with the hands-on stream restoration work, these young people attend classes in conservation subjects such as aquaculture, environmental awareness, erosion control, and fire fighting. There they learn job skills for the future and in many cases meet basic education requirements for high school diplomas.

The Salmon Restoration Project (SRP) was started in 1980 by legislative act SB 201. The program took CCC members from the Humboldt Center at Weott and put them to work removing logjams

Massive logjam blocks a stream. When such logging debris is being cleared, some material is left undisturbed to help restore natural rearing habitat. (California Trout, Inc.)

or modifying those jams and other barriers to migrating steelhead and salmon. The initial goal was to clear one hundred miles of stream each year. With most known barriers removed, the group's focus has turned toward stream restoration and enhancement. Their record is impressive: to date they have cleared or improved more than five hundred miles of tributary streams of the Eel, Mattole, Van Duzen, and South Fork Trinity rivers, along with tributaries to Humboldt Bay and various Mendocino coastal streams. Most important, salmon and steelhead are now spawning and rearing in waters where they had not been seen during this century.

The present SRP is funded annually by a million-dollar contract between the CCC and the Department of Fish and Game. Humboldt Fire Center is the base location for the project, with satellite crews in Trinity and Mendocino counties. A total of sixty corps members, augmented occasionally by other youth from the Fire

Center and a Eureka nonresidential satellite, are involved in SRP activities.

These crews, under the direction of a DFG project leader, use the latest state-of-the-art techniques and equipment to restore and improve fish habitat. Project leader Gary Flosi, through contacts with a wide range of community leaders, landowners, and industry representatives, sets up projects. Seeking advice from independent contractors and others, he works out schedules, provides training, and maintains quality control for the CCC's habitat restoration crews. With technical help from the DFG, these crews learn vital "how-to" skills of bank protection, water deflection, gravel retention, and pool development. They know they've learned their job when they see the first salmon or steelhead utilizing the habitat they have prepared for them.

During fiscal year 1987–88, SRP workers surveyed and developed work plans for sixty-four streams. Fifty-eight fish migration barriers were modifed or removed. (Since woody debris is a natural part of a streambed, habitat restoration workers avoid "clearing" a stream more than is necessary for fish passage. Within riparian zones and on upslopes, eighty-five thousand trees, representing eight indigenous species, were planted. Three thousand feet of streambank was stabilized with log cribbing and riprap. One hundred and three instream structures for pool development and spawning gravel retention were constructed, including log and combination boulder/log weirs and wing deflectors. Eighty-four log cover structures were placed to provide sheltered pool habitat. Altogether, about thirty-eight miles of stream and streamside habitat were improved in thirty-six watersheds during that one year.

The young men and women of the CCC come from a variety of locations throughout the state and from other states. They enroll for a year, with an additional year optional, to do a variety of fish habitat work and bank stabilization. They join the CCC for a variety of reasons: inability to find a job, desire to further their education, and simply to get themselves off the street. The desire to find new directions for their lives is commonly expressed. As corps members, they have opportunities to live and learn with other people their own age and come out better at the end of their employment than when they joined.

Fixing a stream. California Conservation Corps workers are restoring a streambank of a Mattole River tributary that had been damaged by poor land-use practice. (Alan Lufkin)

They enjoy not just the learning aspects of the job, but the hard work as well. Says corps member Keith Landers: "Physically, it is a good program. It teaches an uneducated mind to respect and try to preserve the land we live on. It gets the weak at heart and body used to hard work and rugged working conditions." An observer, seeing a crew knee-deep in icy water last January, digging out and hand-winching a log up on a streambank, commented: "Some people are forever complaining that today's kids don't know how to work—I wish they could see this!"

Corps members in the SRP also helped local salmon and steelhead restoration groups in dozens of tasks from construction of rearing ponds to releasing fry into streams. Groups such as the Garberville Rotary Club, College of the Redwoods, the Humboldt Fish Action Council, and the Pacific Lumber Company all utilize help from these young people. R. J. ("Buck") Pierce, Aquaculture/ Fisheries department chair at College of the Redwoods, near Eureka, enthusiastically reports that "many vital goals of our fisheries program (and the campus at large) would never have been achieved without the efforts of the California Conservation Corps."

Another tribute comes from Robert Stevens, forest manager for the Pacific Lumber Company, Scotia, who congratulated the corps and DFG personnel for "their professional attitudes and conduct in all our relations" and assured them of TPL's ongoing cooperation.

A string of accomplishments over the years, the building of ever stronger community cooperation—and especially more fish in the streams—assures that the Salmon Restoration Project will be around for a long time.

Chapter Twenty-seven

For the Sake of Salmon
William Poole

From the ridges above, the lower Mattole River valley is a bucolic scene. Over the millennia, the river has laid a broad green carpet in the middle of this radically up-and-down corner of the state—a carpet now transected by fences, dotted with scattered houses, and edged by open rolling hills. Many of the open hills were once thickly wooded. Old-time ranchers used to cut the big firs and leave them to rot and would sometimes stuff enormous tractor tires full of flaming rags and gasoline and send them on fiery rides down the dry slopes, burning the country again and again to keep it in grassland for their cattle and sheep.

It is only one way in which men have used and misused this land over the last hundred years. First came the tan oak harvest, with thousands of trees ripped from the woods, dragged out to the river's mouth, and shipped down the coast to the tanneries, where the bark was used for curing hides. Then came the cut-and-burn ranchers and, after World War II, the lumbermen, who leveled the old-growth Douglas firs to help fuel the building boom around San Francisco Bay.

By the time the first young urban refugees arrived in the mid-1970s, the exquisite look of this place seemed to many a cruel disguise. The evidence was in the waters, they said. The Mattole and its tributaries, once the clear-water spawning grounds for tens

This essay originally appeared in *This World,* a section of the *San Francisco Examiner & Chronicle,* June 28, 1987; reprinted by permission. © San Francisco Chronicle, 1987.

of thousands of migrating salmon, were running thickly muddy, bank to bank, all winter long. It was the watershed itself that was migrating now—three hundred square miles of country moving gradually out to sea.

The house stands on Lighthouse Road, about halfway between the little village of Petrolia and the mouth of the Mattole River. It was originally built as a Coast Guard barracks down by the beach; a local rancher dragged the building to its present river-bend location sometime after World War II. Rex Rathbun and his wife, Ruth, bought the place in the mid-1970s, after the big ranches along the lower Mattole were split up and sold off to "the new people," folks in retreat from the cities down the coast.

Rex had been an engineer and builder in Marin County, and the couple were twenty years older, with more resources, than many of their back-to-the-land neighbors. "The Ranch" (as the house is still called) was at the end of the telephone and electric lines and quickly became a local communications post and community center. To this day, the Rathbuns take phone messages for half a dozen of their backwoods friends, hanging signs in the kitchen window—a pink square of construction paper for one family, a blue square for another—so folks driving by on the road will know they've had a call.

From the outside the place looks like the main building of a small summer camp. The front yard is big enough to provide dusty parking for a dozen cars, with enough room left over for a full-sized volleyball court. Inside, the house has a large, sunny kitchen and a modest living room paneled entirely in redwood. There's a fireplace, a grand piano, a wall of books, and a pretty woman named Jan Morrison, who sits and braids her sun-streaked hair while Rex Rathbun plays me the call of the spotted owl on his stereo.

Jan Morrison is chairperson of the Mattole Restoration Council, an umbrella organization of groups and individuals dedicated to restoring the Mattole watershed. She is also my unofficial tour guide this day. It was she who had brought me to meet Rex Rathbun, because I told her I am as much interested in people as I am in the land and the river and the fish, and because Rex has been in on the watershed project from the beginning.

Jan has also gathered other council members: Randy Stemler—a slight man with friendly brown eyes—and a bearded fishery biologist named Gary Peterson, whose hair tumbles in a lush ponytail down his back. Rex Rathbun has a beard as well (it is as white as a fair-weather cloud), and his tall, angular frame is dressed in a red T-shirt with a picture of a spotted owl on the front. Rex is hyped up about spotted owls lately and has been going out in search of them in the old-growth timber, which is the only place the birds can make their home.

The movement to save the Mattole fishery sprang up in the late 1970s, leaping to life in half a dozen places up and down the watershed, wherever ecological consciousness had reached critical mass. One such spot is just west of the Rathbun house, where the waters of Mill Creek splash through a culvert under Lighthouse Road and down a little hill into the Mattole River. The culvert probably went in 1964, Rex says, after a great storm washed out the little bridge that used to cross the creek here. When Rex arrived a dozen years later, he noticed the road crew had left a ten-foot fall at the downstream end of the culvert, which meant a nearly impossible ten-foot jump for the silver salmon that should be spawning in the creek. It was a waste, Rex thought. Flowing out of one of the few old-growth stands of Douglas fir left in Mattole Valley, Mill Creek was a translucent winter stream that should have supported many fish. It's the sort of thing you notice when you care about the place you live. It's the sort of thing people were noticing all over the Mattole watershed in those years.

The stream below the culvert is now a jumble of rocks offering a natural stairway to returning salmon. It has been this way since 1979, when Rex and his Mill Creek neighbors convinced the county and the California Department of Fish and Game that the spot should be fixed. Within a few years, Gary Peterson and two other valley residents had formed the Salmon Support Group, capturing adult fish on the river each winter, hatching the eggs, and raising the salmon fry in a homemade apparatus nestled in the trees along the creek, several hundred yards up a steep, brushy trail from the culvert. There is a fifty-five-gallon drum, a chunky hatchbox partly filled with gravel, and a long metal trough for raising the young fish—all strung together with plastic pipe. Because of the silt-free gravel in the hatchbox, the salmon eggs are able to get the oxygen they need to survive. Ninety to ninety-five percent of the eggs taken from adult fish eventually go back into the stream as salmon, compared with a thirty percent survival rate in the muddy river.

In all, the Salmon Support Group operates four hatchboxes in the Mattole watershed (the other three raise the larger king salmon) and has released more than one hundred fifty thousand fish in the last seven years. Some Mattole salmon have been caught as far away as Alaska; others have returned to the river to spawn themselves.

Other efforts throughout the watershed have gone into cleaning up the streams and the river so that natural spawning is more efficient. Banks have been stabilized with hundreds of yards of rock armoring; eroded streamsides are planted with coyote bush or some other rooty native plant to hold the soil in place. Log weirs have been built across streams to collect and retain natural spawning gravel. Political efforts are also under way to secure and preserve what tracts of old-growth timber remain, to discourage the gouging of the land, the muddying of the waters.

What's happening on the Mattole is a model for appropriate fisheries, Gary Peterson believes. The whole hatchbox installation above Mill Creek was built for less than $1,000, including a $35 automatic fish feeder Peterson built from a timer, a length of one-and-a-half-inch dowel, a handful of eightpenny nails, and some brown plastic medicine bottles. The timer turns the dowel through which the nails are driven; the nails tip over the medicine bottles and feed the fish. There are no pumps here, no snaking electrical cables. The water flows into the hatchboxes by gravity and out again into the creek. "This is how things should be," Peterson says. "A small-scale, low-technology, low-cost, site-specific facility, instead of some centralized monument to civilization."

Like most fishery biologists, Gary Peterson had three career options after graduating from Humboldt State University a few years ago. He could have joined the California Department of Fish and Game or the U.S. Fish and Wildlife Service or gone into private consulting. Instead, he adopted a river of his own, moved to its banks (he rents a fifty-dollar-a-month trailer from Rex Rathbun), and assembled a tentative income out of the California Fish and Game contracts and foundation grants that help with the restoration project. When I ask him whom he works for, it takes him a minute to find an answer he likes. "I'm a free-lance fish biologist," he finally says. "I work for the fish."

Peterson is only one example of the inspired blending of vocation and avocation that seems to power the Mattole project. Ask these people what they "do" and the answers are vague and unfocused. A little of this, a little of that; they do what they need to get by. One soon sees that they think of themselves first as residents of the Mattole Valley, subject to all the joys and responsibilities attendant to that status. Their work is play; the play is work. What they do for a living is not as important as living where they do.

Rex is retired and devotes almost all his time now to various community projects. He is an emergency medical technician and a member of the volunteer fire department (both he and Gary carry fire department beepers on their belts) and is actively involved with a group raising money to purchase the old-growth Douglas fir forest that guarantees the pristine flow of Mill Creek.

Randy Stemler and Jan Morrison have learned to grow many of the native plants used for erosion control, and they market plants and seed to organizations and individuals doing restoration work. Stemler has built a battery-powered seed harvester that he wears like a backpack, sucking the seeds off coyote bush by the dozens per minute. Coyote bush, also called greasewood, is ideal for land reclamation, says Stemler, because it likes hot, open sites, forms deep roots, and grows vigorously in impoverished mineral soils.

"I'm sorry I have to come on a weekday," I had told Jan Morrison when I called her from the Honeydew store that morning.

"Oh, don't worry," she had said. "That's not such a big deal out here."

Late in the afternoon, Gary, Jan, and Randy drive me to a high point overlooking the mouth of the river. Away to the west, the dark ocean stretches to the horizon. To the north, Cape Mendocino, the westernmost point in the country, points the way to sea. It is early June after a dry winter, and the Mattole has lost the force it needs to reach the ocean, its waters backing up the estuary behind a narrow sandbar. It rains infrequently here at this time of the year, but high clouds are scudding in from the south, and Gary Peterson hopes a late storm might open the mouth of the river for another day or two to let some of the smolts out to sea.

The estuary is one of Gary's big concerns these days. Before the logging and burning and grazing, the river was a liquid ribbon beneath a green riparian canopy, and old-timers still talk of holes so deep nobody could plumb their bottoms. Salmon thrived in the cool deep water, on a rich diet of aquatic insects hatched in streamside plants, and went to sea later, stronger than they do now. Today the estuary is a bland lake, silted up, baking in the open sun, and Gary Peterson is already trying to think of ways to fix it, planting trees along the north bank, fencing the cows from the green slough along the south bank, sinking logs or tangles of tree roots to provide shelter for oversummering fish.

"We've always had this economy based on extracting resources," says Jan Morrison. "I think we've come to a point where we have to find economies that put things back, that nourish the land."

In the Mattole Valley, it is the salmon that will finally judge the success or failure of such efforts. Salmon, like spotted owls, are indicator species, Rex Rathburn had reminded us that morning—species sensitive to the least degradations of the ecosystem. Spotted owls are trusting creatures who can live only in dense, old-growth woodlands. Salmon can thrive only in the clearest, healthiest streams.

Land where such creatures live is wholesome from the inside out, fit for habitation by other living things, including the large bipedal mammal *Homo sapiens*, a species that is occasionally as skilled at life, growth, and renewal as it is at death, as clever at caring for the land as in using it up and throwing it away.

Chapter Twenty-eight

Urban Stream Restoration
William K. Hooper, Jr.

For years, small urban streams have been degraded and neglected, causing a loss of their fishery resources. Recently, however, privately organized and funded local restoration efforts have been gaining momentum statewide. These efforts may provide the answer to restoring salmon and steelhead fisheries in urban areas.

Restoration opportunities are many. In the San Francisco Bay area, there are more than fifty-seven streams—not counting the Sacramento and San Joaquin rivers—fed by more than one hundred smaller tributaries. Historically, virtually all of these streams supported spawning runs of salmon and steelhead. Today, nearly all of these streams suffer from the familiar ills of urbanization and poor resource management: water diversions, channelization, instream barriers, sedimentation, destruction of riparian growth, pollution.

Only a few Bay Area streams, estimated to be less than 5 percent, could provide the habitat necessary to support salmon and steelhead populations. These streams include Napa River, Petaluma River, and San Jose's Guadalupe River, as well as Corte Madera, Sonoma, Walnut, Wildcat, Alameda, and San Francisquito creeks. Among the hundred and sixty-six streams in the Bay Area, these nine streams seem to offer the best potential for restoration.

Surprisingly, these streams have remnant salmon and steelhead runs and only partially degraded habitat. While proof is not available, it is estimated that the number of spawning fish in these creeks has declined to less than 5 percent of historical levels and is continuing to fall. Alameda Creek, for example, once hosted runs of

Seasonal dams with no fish protection facilities, like this one at Healdsburg, harm both upmigrant and downmigrant salmon and steelhead. (Stan Griffin)

more than two thousand steelhead. Today probably fewer than one hundred fish can be found there.

It is unlikely that these streams could be restored to support traditional fishing pressure. Many could, however, be restored sufficiently to serve as urban preserves or outdoor classrooms—special places where one can glimpse the magnificent cycle of salmon and steelhead. These urban watersheds could also stimulate increased public awareness of problems facing fishery resources statewide.

Local private efforts with help from state and federal resource agencies appear to offer the best and only apparent hope for restoration. Urban streams are typically too small to merit sufficient attention from the California Department of Fish and Game, the primary agency responsible for protecting urban watersheds. To date, where urban streams have been restored, a privately organized local group has been the spearhead. Generally, the CDFG provides counsel and expertise but little in funds. Another agency, the California Department of Water Resources, offers seed money from its Urban Streams Restoration Program for local stream projects. But manpower to restore streams is not available from CDWR.

The record of local citizen groups in marshaling resources is encouraging. One inspiring project is conducted on the Petaluma River in Sonoma County, where students of Petaluma's Casa Grande High School, under the direction of natural resources instructor Tom Furrer, have brought salmon and steelhead conditions back to life. During the last three years, students removed more than twenty truckloads of debris from critical spawning areas. They planted five hundred redwood trees along degraded streambanks. They established a daily "fish watch" to monitor the number of spawning fish in the watershed. In 1988 they released twenty-seven hundred steelhead smolts in the Adobe Creek tributary. They are currently building a hatchery with a production capacity of five hundred thousand steelhead and thirty thousand striped bass.

On San Francisquito Creek in Palo Alto, a group supported by California Trout, Inc., is making headway in removing barriers and diversions and studying hatchery and increased water flow options on the creek.

In Santa Cruz, over the last thirteen years, salmon and steelhead have been restored in the San Lorenzo River, which runs through the downtown and famous boardwalk area. A very impressive local team of fishermen and biologists has dramatically reversed the loss of salmon and steelhead. The group, a nonprofit organization called the Monterey Bay Salmon and Trout Project, has a long list of accomplishments, including building and operating a hatchery and rearing facility on nearby Big Creek. The hatchery has a higher production rate than most state and federal hatcheries at a fraction of the cost. All this has been accomplished with private funds.

Wildcat Creek in Richmond is a live stream today only because nearby residents refused to accept a conventional flood control project that would have devastated the stream. Controversy over that flood control project helped educate engineers and laypersons alike to the physical, fiscal, and social advantages of nonstructural solutions to stream problems.

Alameda Creek, for several years the subject of seasonal damming for flood control and water storage, also has attracted attention of local residents. A private group, the Alameda Steelhead

Restoration Committee, has done a superlative job of identifying possible solutions to problems of restoring the fishery. Seasonal dams in lower reaches and unfavorable water rights still pose serious problems, however.

Corte Madera Creek in Marin County benefited from a project organized by members of a fishermen's group, Trout Unlimited, who contributed labor and materials to rebuild a fish ladder on the creek. This significantly improved salmon and steelhead viability on the stream. However, stream conditions require further improvement, such as obtaining increased water flows year-round from the Marin Municipal Water District and controlling runoff and sedimentation in the upper watershed.

On the education front, several school districts in the Eureka-Arcata area have developed elementary school curriculums that include classroom aquarium hatching facilities and field trips to local streams. These programs are integrated into science courses, mathematics, and social studies projects focused on citizen activity to help solve environmental problems. Diane Higgins, a local teacher and fishery advocate, developed a course of study in this area under the aegis of the California Advisory Committee on Salmon and Steelhead Trout. Jeff Self, curriculum coordinator with the Eureka City Schools system, has established a mini-hatchery at Washington Elementary School and conducts teacher in-service preparation classes. Much of the funding for these projects is derived from sales of personalized auto license plates (Environmental License Plate Fund) and local and statewide fishery groups.

Schools elsewhere in California are conducting fish conservation programs as well. In Redding, middle school teacher Kathy Callan has established a thriving "Adopt-a-Stream" program involving her seventh graders that receives much support from parents and local fish and wildlife professionals. In the Monterey Bay area, educators Barry Burt and Matt McCaslin are developing and disseminating classroom education programs. Connie Ryan, with the U.C. Cooperative Extension/Sea Grant program in San Francisco, is introducing fishery restoration education programs in the Bay Area. Elena Scofield, of the Department of Fish and Game, is heading efforts to establish "Project WILD," a nationwide natural resource education program, in the California public school system.

These citizen projects demonstrate that urban stream restoration can work, but they require private initiative combined with assistance and expertise from state and federal fishery and water agencies. These local, privately spearheaded efforts appear to offer the most effective and productive hope for restoring urban stream fishery resources.

Chapter Twenty-nine

Saving the Steelhead
Eric Hoffman

Retired businessman Dick Wehner is an avid fisherman and is known in Santa Cruz fishing circles for being as good as they come at coaxing a salmon or steelhead into taking his lure in the San Lorenzo River. But these days anybody looking for Wehner is more apt to find him counting steelhead eggs or pitching in to build hatchery ponds at the Kingfisher Flat Steelhead Fish Hatchery tucked away in a rugged coastal canyon north of Santa Cruz. It's not that Wehner's fish-finding faculties have diminished. They're as good as they ever were, maybe even better than the day in 1982 when he finessed a fifteen-pound buck steelhead on light tackle from the San Lorenzo River.

Recently Wehner waded into Big Creek near Davenport to show his fisherman buddy, Pat Totaro, an extremely rare fish, a coho salmon. Once plentiful in the streams north of Santa Cruz, this was one of only three native males to return to Big Creek this winter to fulfill its final mission of propagation. To the disappointment of Wehner and others, there were no females. The male Wehner flushed out by prodding the underside of a snag with a stick is, at first glance, a thirty-inch streak of silver and pink, but when it pauses you can see the telltale white and gray splotches that signal it has only a short time to live. The fish will die according to nature's clock, but with its mission unfulfilled. Native salmon here

This essay was first published in *Monterey Life*, March 1986; reprinted with permission.

are probably—like the California condor—doomed, and the male fish Wehner located may be the last vestige of ageless salmon runs that have steadily petered out during the last thirty years in all the rivers and streams that feed the Monterey Bay.

By all accounts, the future of native salmon looks bleak. But thanks to the efforts of a small group of private citizens, the local populations of steelhead trout, their anadromous cousins—fish that are born in fresh water, live in the sea, and return to their freshwater birthplaces to spawn—stand a good chance of being saved. Locally, anadromous species are coho salmon and steelhead.

Greater breeding longevity in steelhead may account for their greater numbers during adversity, while native salmon have all but perished. Just how well the steelhead will do depends on a combination of sporadic funding, support from fishermen's associations, dedicated scientific know-how, and always-fickle natural elements.

Due largely to clever manipulation of what fish remain, and plenty of hard work, steelhead are increasing in numbers. The project has steadily picked up steam since its inception in 1975. The diligence of the project biologist Dave Streig, volunteer work efforts, donations of materials and land, have been supplemented with a $16,000 grant from the Packard Foundation. It has all added up to a dramatic reversal of the downward trends in steelhead populations.

About fifty volunteers take turns driving up Highway 1 to Swanton, past the turn-of-the-century McCrary homestead, to the Kingfisher Flat Fish Hatchery. The McCrarys, whose Swanton ancestry dates back to the 1840s, donated the land and materials for the project, officially dubbed the Monterey Bay Salmon and Trout Project. The volunteers do everything from counting the eggs of freshly milked hen steelhead to constructing ponds. All told, the Monterey Bay Salmon and Trout Project has put over three hundred fifty thousand fish in the San Lorenzo and Pajaro river systems and many of the creeks surrounding the Monterey Bay.

Lud McCrary likes to talk about the efforts to save steelhead on his family's land. His perspective and goal for the fish would get a nod of approval from everyone involved in the project. McCrary: "I'd like to see the population back to where it was when I was a kid, hopefully more than when I was a kid."

McCrary can recount childhood memories of the 1930s. During

the Depression his uncle poached salmon at night with a long pole and gaff hook by "feeling" the fish and plucking them from the local streams. His uncle fished for pragmatic reasons: to feed his family. He fished at night for other reasons: to avoid McCrary's grandfather who was a part-time warden in the area. McCrary can also remember a rich fish smell from rotting flesh of spent salmon that filled the deep, quiet redwood canyons along the North Coast, but not since the 1950s.

It's not known for sure what forces are responsible for the hard times experienced by the salmon and steelhead locally. The list of possible culprits is long. Commercial fishing operations may have had an impact on salmon. Silting and alteration of river systems due to human activities is a probable factor. Toxicity problems that have plagued so many marine life forms can't be overlooked. And long-term climatic changes, punctuated with events like El Niño, may have played a part.

Nobody knows for sure because no studies were undertaken until recently, and the salmon are essentially gone. Trying to make sense of what has gone on historically in terms of numbers is nearly impossible. Even though the California Department of Fish and Game has operated along the Central Coast since the 1930s, their records of fish counts have been lost.

Where bureaucratic paper shuffling accomplished little, the perseverance of citizen-fishermen like Dick Wehner and Pat Totaro seems to work. The program receives less than 50 percent of its funding from state sources. The fish have plenty of boosters from the private sector. Superior Court Judge Bill Kelsay and former Santa Cruz Supervisor Dale Dawson are among the many who have put in countless hours constructing ponds and collecting donations for their cause.

But if fish could talk they'd probably point to Jack Harrell, Dave Streig, and the McCrarys as the most essential cogs in their revitalized waterwheel. Not only did the McCrarys supply the land and materials, they also kept Big Creek in a near-pristine state, supplying the hatchery with a ready source of native fish. Jack Harrell, whose paid job encompasses all aspects of maintenance of the Santa Cruz Wharf, has spent thousands of volunteer hours rebuilding check dams, stringing nets above breeding ponds to keep herons from swiping the precious steelhead, and overseeing construction of a new hatchery building.

But all the efforts are for naught without a healthy dose of good management, and that responsibility falls to biologist Streig, the project's only paid employee. After spending a few minutes with the affable Streig, it is clear he is doing more than just a job. His ancestry dates back six generations in the Pajaro Valley. Like the McCrarys, Streig remembers stories from past generations when salmon and steelhead were plentiful. He thrives on the success of the fish he nurtures. "It makes me feel good to see the fish we've tagged return. That's why winter is my favorite season at the hatchery. Last year over 50 percent of the fish caught in the San Lorenzo were from the Kingfisher Flat Fish Hatchery, and this year the number will be over 75 percent. I think we can be proud of that."

A great deal of Streig's success is due to his understanding of the subtleties of managing local fish populations. Unlike the California Department of Fish and Game, which has generally operated under the assumption that all salmon and steelhead are nearly the same, Streig sees things differently. While Fish and Game tackled steelhead restoration projects throughout the state with a single centralized hatchery on the Mad River in Humboldt County, Streig sees great value in a regional approach. Interestingly, the state's efforts at restocking worked well in the northern part of the state, but not so well on the Central Coast.

It's hard to argue with success. Streig's emphasis on a regional approach, one that develops native strains rather than relying on nonnative forms, has worked well. "Apparently our native populations have subtle survival adaptations that allow them to cope with higher water temperatures and adjust to unique aspects of our river and stream systems," he says. "Fish of nonnative genetic origins don't do very well." Additionally, planting fish from outside the area may even harm struggling local populations. "When fish that don't do as well breed with our local population, they weaken the genetic makeup of the local fish and hurt their ability to survive," Streig adds. With this reasoning, he has developed a regional hatchery approach that gathers eggs from existing fish, safely guards them, and raises them to ensure that the maximum number of fish survive.

The actual work of creating tens of thousands of fish is, in two words, tedious and unending. Steelhead and salmon are among the most difficult fish to raise in an artificial setting. The water temperature must be kept low because if it rises in a hot spell, the fish weaken and are susceptible to any number of diseases that can kill off a whole season's work in a few days. "That's why the Big Creek site is so good," Streig explains. "The water temperature usually stays between forty-five and fifty-eight degrees, which is optimum for these fish." But when hot spells sweep through the area, the respiration rate of the crowded fish rises, quickly depleting the oxygen supply needed in vast amounts in order for them to survive. Streig then revs up the immense generators that pump air through the ponds. "It's a never-ending task when people assume the role of Mother Nature."

Streig, who can be found at the hatchery from five to seven days a week, has had some nervous moments with his slippery charges. He came to work one morning and found forty piles of young salmon and great blue heron tracks around the hatchery ponds.

In the middle of winter when many people would find the tree-shrouded, mossy canyons like Big Creek too dank and depressing, Streig, in his rubber wader work attire, is all smiles. "This is what it's all about. There's water in the streams and the fish that we sent to sea years earlier come home so we can help them lay their eggs. Just yesterday I milked hen #119 and got thousands of healthy eggs. She grew from twenty-seven to twenty-nine inches and increased her weight from eight and three-forths to ten and one-half pounds during the last year." Streig put her back into Big Creek and hopes she'll return next year. In the meantime, he has tens of thousands of eggs to watch over.

Summary and Conclusions

The contributors to this volume have shown that a valuable California natural resource, its depleted salmon and steelhead populations, is in danger of being lost. Prehistoric populations of these fishes undoubtedly numbered in the tens of millions. At present estimates, adult populations of chinook salmon may be as few as one million adult fish in a given year; coho salmon numbers are one hundred thousand; steelhead total perhaps two hundred fifty thousand fish. At least one California salmon species is extinct. The Sacramento River winter-run chinook is endangered; another Sacramento River species is in very serious decline.

A major recent development, however, suggests reason for optimism: in 1988, the California legislature enacted SB 2261, the Salmon, Steelhead Trout, and Anadromous Fisheries Program Act. This law, conceived by the California Advisory Committee on Salmon and Steelhead Trout, established a statewide goal of doubling stocks of these fishes by the year 2000—primarily by restoration of habitat and maintenance of natural stocks. The act details actions essential to realization of that goal based on the Advisory Committee's detailed four-year study, which involved hundreds of Californians.

Is such a goal realistic? The obstacles to its realization are formidable. The first obstacle is the lack of adequate amounts of suitable water to maintain healthy fish habitat. Until recently, fishery interests have never had sufficient political clout in state government or the courts to contest successfully for needed water.

Agriculture, notably modern agribusiness, has long dominated California water development and allocation. Its working partners have included federal agencies, a compliant, politically vulnerable state water regulatory structure, and historic precedent. Water rights are protected zealously. The largest diverter of Sacramento River water, the Glenn-Colusa Irrigation District, on a bronze plaque displayed at their main pumping plant near Hamilton City, cites biblical authority: "Thou shalt take of the water of the river and pour it upon the dry land" (Exodus 4:9).

Municipal water districts are another key factor in California water development and allocation. Their impact, even if agricultural irrigation were not a factor, could be overwhelming to fishery restoration efforts. The state's population is increasing at an annual rate of 2.4 percent. Demands for water are greatest in arid, high-growth southern California areas, where homes must be provided with water and dependable water supplies help to create markets for still more homes.

Because California irrigation ditches and urban faucets have always produced water on demand, the influential bloc of agricultural area and urban voters has had no reason to be concerned about the competing water needs of salmon and steelhead—fish that exist for the most part in cool northern waters.

Several developments in recent decades suggest, however, that this situation may not continue indefinitely. The placement of key California rivers in the state (1972) and national (1981) Wild and Scenic Rivers systems, and voters' resounding defeat of the Peripheral Canal plan (1982), were favorable to fish and wildlife interests. More recently, a welter of federal and state legislation addressing public environmental concerns, strong voter support of state and local park and wildlife propositions, and results of formal polls signal growing public sentiment favoring protection and restoration of natural resources wild, clean, and pristine. Court decisions during the 1980s have raised public trust questions regarding water rights and water quality issues. The State Water Resources Control Board must take public trust issues into account as it makes allocation and water quality decisions, as exemplified by the current effort to establish Bay/Delta water quality standards, which will affect water rights.

Issues related to water economics in California may also affect allocation of water supplies to meet the needs of agriculture, urban water districts, and fisheries. Governmental policies that provide publicly owned water at far below cost to corporate farms that produce surplus, subsidized crops, and pollute the water in the process, are seriously questioned. This raises the possibility of water "deregulation." Establishment of a free market in California water, claim its supporters, would undoubtedly boost its price and necessitate stricter controls on its offstream allocation and use. (Fish activists insist, however, that public funds should never be used to buy water for fish, because both the water and the fish are public property.)

Other water-based issues affect the agriculture and water industries and may diminish their political dominance. Policymakers must consider such realities as farm soils made unusable by accumulations of salt from irrigation waters, downstream effects of flushing of natural toxic materials (such as selenium) and chemical pesticides into waterways, and saltwater intrusion into aquifers.

More positive forces are also at work, albeit on a small scale at present. Alternative solutions for farmers' water problems are being explored in universities and government. For example, the U.S. Department of Agriculture is currently studying proposals for development of LISA—low-input sustainable agriculture—to reduce dependence upon "certain kinds of purchased inputs," including water. Statewide workshops to improve irrigation efficiency are sponsored by California Polytechnic Institute. Since the 1976–1977 and more recent droughts, agricultural and urban water districts have studied and implemented other water conservation measures. (It is often reported that agriculture accounts for the use of 85 percent of the state's developed water. Spokespersons for agriculture claim that 70 percent is a more accurate figure today.)

Some developments, however, which appear favorable toward meeting objectives of SB 2261, such as provisions for greater public involvement in water allocation decision-making processes or claims of improved relations between "water folk and fish folk," are magical acts, say fish activists: "While you're focusing on one hand, the other is deftly picking your pocket." Mistrust dies hard.

North Coast habitat problems of a different sort constitute a second obstacle. Although insufficient water is a serious concern in some northern California communities, rainfall, ironically, contributes to other problems: because of the region's heavy precipitation and highly erodible soils, logging-related activities cause immediate runoff that destroys natural streams and their biosystems, often for miles below the initial perturbation. Since lumbering has traditionally been the region's principal bread-and-butter industry, protection of habitat for salmonid resources has not been a compelling consideration.

But there are also hopeful signs for fishery restoration on the North Coast. The lumber industry is coming to recognize that environmental conditions suitable for growing salmon and growing trees are remarkably similar, and fish activists are gaining support. Leadership from local educators, commercial fishery leaders, and clusters of community activists, with growing governmental, lumber industry, and landowner support, is improving the climate for fishery restoration. Today, salmon and steelhead restoration is becoming a popular element of science, mathematics, and social studies curricula in many North Coast public schools.

Although obstacles exist, statewide restoration of salmon and steelhead stocks is clearly possible but certainly not assured. The economic, political, and social forces that caused the decline will not change soon. Moreover, although fish activists may find reasons for encouragement, changes in the agriculture, water, or lumber industries will not automatically translate to benefits for fish or other wildlife resources.

This suggests a third obstacle: public attitudes. Complex ecological concepts are not widely understood. Environmentalism, particularly after the shock treatment of several major oil spills in recent years, is becoming politically fashionable. But until recently, most Californians showed only limited concern about resource conservation issues. Our cornucopia of natural resources—clean air, sweet water, healthy soils, myriad life forms and their complex interrelationships, the *ecology* that includes humans—has customarily been accepted, and exploited, without serious examination.

Most citizens have not been concerned about depletion of resources because they have not experienced the problem. When such issues begin to become apparent, this nonconcern must be penetrated before substantive problems can be resolved. This requires large infusions of funds to redirect public opinion. Environmental groups typically lack funding, while interests happy with the status quo enjoy the dual advantage of having more money and less need to change people's attitudes.

Suggested compromises for resolving salmon and steelhead resource problems have surfaced. A scenario encouraged by the water industry would include downplaying the importance of San Joaquin River restoration and assigning a lesser role to the Sacramento/San Joaquin Delta for downstream migration. Young fish produced by improved and expanded hatchery and fish passage facilities on the Sacramento River would be released almost exclusively in the San Francisco Bay-Estuary. North Coast streams, where restoration projects appear to flourish, would become a more important base for production of salmonids for commercial and sport harvest. Agribusiness would thus benefit from freer access to Sacramento and San Joaquin river water.

Such a compromise is unacceptable to fishery experts and the Advisory Committee, who insist that the only solution to California's salmon and steelhead problems is overall restoration of habitat and maintenance of natural stocks. The ecological health of the entire San Francisco Bay/Delta ecosystem is considered essential to resolution of statewide fishery problems.

Moreover, each side in a compromise is expected to give up some benefit. Fishery interests feel the resource has already been "compromised" nearly to death. Further deterioration of already decimated habitat and decline of natural stocks are untenable, given the severely depressed state of the fisheries.

At the policy level, outcomes of the current State Water Resources Control Board's Bay/Delta water quality hearings are crucial to improvement of fisheries. Resolution of Central Valley Project water contract problems is an immediate issue. A coalition of environmental and fishery groups is contesting renewal of Friant Dam water contracts because the Bureau of Reclamation proposes to renew those contracts without environmental impact studies.

The initial question remains: Is the Advisory Committee's goal to double salmon and steelhead stocks realistic? Early in the twenty-first century, will salmon and steelhead numbers actually be twice what they are now? Can the obstacles to meeting this major goal of SB 2261 be overcome? The restoration effort required will be immense. The continuing action of pro-fishery interests, capitalizing on favorable court decisions and emergent cooperation from government and industry, is essential.

Says a prominent fishery activist: *"People lead, and government follows. Public insistence—informed, focused anger—favoring restoration is a necessary prerequisite for progress."* Without such a commitment, other efforts cannot avert the ultimate destruction of California's salmon and steelhead resource.

Abbreviations

ACID	Anderson-Cottonwood Irrigation District
AFS	American Fisheries Society
BIA	Bureau of Indian Affairs
BuRec	Bureau of Reclamation
CCC	California Conservation Corps
CEQ	Council on Environmental Quality
COA	Coordinated Operation Agreement
CRMP	Cooperative Resource Management Plan
CVP	Central Valley Project
DFG, CDFG	California Department of Fish and Game
DWR, CDWR	California Department of Water Resources
EIR	Environmental Impact Report
EIS	Environmental Impact Statement
EPA	Environmental Protection Agency
FERC	Federal Energy Regulatory Commission
FWS, USFWS	United States Fish and Wildlife Service
GCID	Glenn-Colusa Irrigation District
KMZ	Klamath Management Zone
LADWP	Los Angeles Department of Water and Power

NEPA	National Environmental Policy Act
NMFS	National Marine Fisheries Service
NRDC	National Resources Defense Council
NWFF	Noyo Women for Fisheries
PCFWC	Pacific Coast Fishermen's Wives Coalition
PFMC	Pacific Fisheries Management Council
RBDD	Red Bluff Diversion Dam
SRA	Salmon Restoration Association
SRP	Salmon Restoration Project
SWP	State Water Project
SWRCB	State Water Resources Control Board
TCC	Tehama-Colusa Canal
TPL	The Pacific Lumber Company
USFS	United States Forest Service
WAPA	Western Area Power Administration
WPA	Works Progress Administration

Glossary

Adipose fin

Small fleshy fin on the back of a salmonid just ahead of the tail fin. (Often excised to identify fish.)

Aeration gear

Mechanical device to circulate air through water for support of fish life, usually while fish are being transported.

Alevin

The newly hatched salmon or steelhead still in the gravel with its yolk sac attached.

Anadromous

A term describing fish that migrate from salt or brackish water to fresh water (commonly their natal streams) to spawn. Salmon and steelhead are anadromous.

Aquifer

Any geological formation of sufficient porosity and permeability to store, transmit, and yield water to wells and springs.

Armoring

Process by which streambeds, upon loss or cementing of fine materials, become impenetrable to salmonids attempting to dig redds. Also refers to an artificial streambank protection process involving placement of rock or other permanent stabilizing materials.

Beneficial use

A use of water for some economic or social purpose. The State Water Resources Control Board recognizes twenty-one beneficial uses of water and establishes the levels of water quality required for each.

271

Biotic	Relating to life and living systems rather than the physical and chemical characteristics of an environment.
Broodstock	Parents solicited in bioenhancement or artificial propagation efforts.
Buck	Informal term for adult male steelhead.
Bypass	A channel used to divert flows from a mainstream, as for the diversion of floodwaters. Also refers to a required release flow for fish maintenance instream.

California,
State of:

Department of Fish and Game	A department of the Resources Agency charged with administration and enforcement of the Fish and Game Code and carrying out policies formulated by the Fish and Game Commission, the Wildlife Conservation Board, and the Marine Research Committee. The director is appointed by the governor.
Department of Water Resources	This department constructs and operates the State Water Project and is responsible for planning state water resources, regulating dam safety, and controlling floods (using federal criteria).
Fish and Game Code	The body of laws and regulations relating to protection, propagation, and preservation of fish and game in California.
Fish and Game Commission	A body appointed by the governor that formulates general policies to be administered by the Department of Fish and Game. The commission's powers are delegated to it by the legislature.
Marine Research Committee	This committee, appointed by the governor, consists of private citizens representing mainly commercial fish interests. It is empowered to employ personnel or contract for research on commercial fishery resources of the Pacific

Ocean. Its financial support derives from a special tax on fish dealers and fish packers.

Reclamation Board   Part of the Department of Water Resources, the Reclamation Board is the state agency responsible for the development and implementation of flood control plans on the Sacramento and San Joaquin rivers and their tributaries.

Regional Water Quality Control Boards   Nine such boards, representing the state's major basins, issue permits for wastewater discharges to surface and groundwater and administer enforcement of permits.

State Lands Commission   Part of the Resources Agency. Through its staff (the State Lands Division) the commission sets policies for the administration of more than 3.5 million acres of sovereign and public land, primarily tidal and submerged lands, in California.

State Water Commission   A body appointed by the governor to monitor water use in California and recommend state water policies.

State Water Resources Control Board   A body appointed by the governor that controls permits for the use of water, except groundwater and riparian uses. It adopts water quality plans, regulations, and policies.

Wildlife Conservation Board   A group consisting of the president of the Fish and Game Commission, the director of the Department of Fish and Game, the director of finance, and three members each from the Assembly and Senate. The board's primary responsibilities are to select and authorize acquisition of land and property suitable for recreation purposes and the preservation, protection, and restoration of wildlife.

Carp   An exotic species of fish commonly found in Central Valley streams.

Carrying capacity   Equilibrium in population size for a particular species in a given situation, marked by a balance between its reproductive potential and the environmental resistance.

Central Valley Project	A major statewide water development project built and operated principally by the U. S. Bureau of Reclamation.
Chinook salmon	(*Oncorhynchus tshawytscha*). Largest species of Pacific salmon; only species found in Central Valley streams. Found also in larger coastal streams. Also called king salmon.
Coho salmon	(*Oncorhynchus kisutch*). A species of Pacific salmon found most commonly along California's northern and central coast. Also called silver salmon.
Commercial fishing	Fishing for sale to markets.
Conservation	The protection, improvement, and use of natural resources according to principles that will assure their highest economic or social benefit.
Conservationist	A person committed to the rational use of the environment to improve the quality of living for mankind. In practice, synonymous with environmentalist.
Cooperative Resource Management Plan	A plan to deal with cumulative effects of overall land use in which various interested parties jointly develop restoration plans. The concept is finding favor along the North Coast among fishermen, the lumber industry, farmers, and other user groups.
Coordinated Operation Agreement	A 1985 agreement between the United States and the California Department of Water Resources for coordinated operation of the Central Valley Project and the State Water Project. The U.S. Fish and Wildlife Service, in cooperation with the California Department of Fish and Game, recommends policies for protection of fish and wildlife affected by the agreement.
Cubic feet per second	A basic unit for measuring the flow of water past a given point. Equivalent to 449 gallons per minute and 1.98 acre-feet per day.
Decadent	Trees in an advanced state of decay or decline in comparison to an earlier condition of vitality.

	Often used as a technical forestry term where net growth of a tree or stand of trees is negative.
Detritus	Loose material that results directly from the disintegration or abrasion of rock and organic matter.
Diversion	Removing water from a stream by any means that is not natural. Dams, pumps, and conveyance structures are common means of water diversion.
Downmigrants	Juvenile salmon or steelhead migrating toward the ocean, where they may spend most of their lives.
Drainage basin	The land area from which water drains into a stream system. See *Watershed*.
Drawdown	The magnitude of change in surface levels in a well, reservoir, or natural body of water resulting from withdrawals.
Driftnet	A gillnet permitted to drift freely in ocean currents.
Ecosystem	The interacting system of a biological community and its nonliving surroundings.
El Niño	A periodic warm-water ocean event, usually confined to the equatorial Pacific, that occasionally affects eastern Pacific water temperatures as far north as Alaska. It alters distribution of fishes and thus decreases West Coast salmon fisheries for a time.
Endangered Species Act	The federal act (1973) that protects designated species and their habitats; it requires any development to be halted if it can be shown to endanger a species or subspecies throughout all or a significant portion of its range. The California Endangered Species Act (1984) is the state's equivalent measure.
Enhancement	Improvement of existing conditions independent of previous factors that affected fish populations or habitat. Bioenhancement refers to hu-

man activities designed to improve habitat conditions for plant and animal life.

Environmental impact statement (EIS)

A document required of federal agencies by the National Environmental Policy Act for major projects or legislative proposals. An EIS provides information for decision makers on positive and negative effects and lists alternatives to the proposed action, including the option of taking no action. An environmental impact report (EIR), California's version of the EIS, is required under the state's Environmental Quality Act.

Environmentalist

See *Conservationist*

Escapement

The number of salmonids that survive (escape capture) to spawn in freshwater streams.

Estuary

Area where fresh water meets salt water (bays, mouths of rivers, salt marshes, lagoons). This brackish-water ecosystem shelters and feeds marine life, birds, fish, and other wildlife.

Fish Commission

A governmental body (initially national) authorized to study and regulate fisheries. The California Fish Commission of the late 1800s became the Fish and Game Commission.

Fish ladder

In a flowing stream, a series of pools of graduated elevations that permit fish to pass a barrier over which they cannot leap.

Fish screen

A mechanical device that prohibits fish passage while permitting water to flow.

Fisher

Generic term for fisherman, fisherwoman, or social group involved in subsistence or commercial fishing.

Fishery

Fish resources, their habitats, and people who use the resources.

Fishery Conservation and Management Act (Magnuson Act)

A 1976 law that declared federal jurisdiction over fisheries between three miles and two hundred miles offshore and introduced new methods of managing them.

Flashboard structure	A dam with wooden panels that may be added or removed to change levels of water impoundment.
Floodway	See *Bypass*.
Furunculosis	A highly infectious bacterial (*Aramonas salmonicida*) disease frequently characterized by boils or furuncles.
Gaff	A hook, usually two or more inches across, attached to a handle, often used to lift fish from water to a boat. A gaff is commonly used by poachers to catch fish in streams.
Gene pool	The total collection of genes found in a population of organisms.
Genome	In genetics, a single haploid set of chromosomes of any organism. (Haploid means having half the normal number of chromosomes.)
Gillnet	A floating mesh trap that captures fish by ensnaring them behind gill covers as they try to pass through the net.
Grilse	Two-year-old sexually mature salmon, usually male chinooks, returning a year early to streams from the ocean. Commonly called jacks or chub salmon. Grilse are not ordinarily used for artificial propagation.
Gurdy	A powered hoist used on salmon boats to raise and lower troll lines.
Gypo	Small-scale timber harvester common on the North Coast during the 1950s and 1960s. Also called "cut and run" operator.
Habitat	The sum of environmental conditions in a specific place that may be occupied by an organism, population, or community.
Hatchbox	A small-scale method for incubating eggs of wild fish. Hatchboxes are a common element in off-stream restoration projects on North Coast streams.

Headwaters	The upper reaches of a river; the small streams at the upper levels of a watershed.
Hen	Informal term for female steelhead.
Hoopnet	A net attached to a pole and used to scoop fish from a stream.
Hydraulic mining	A formerly common means of gold mining in California in which desired minerals are recovered from gravel deposits dislodged by a powerful jet of water. Banned by the U.S. Circuit Court (the "Sawyer Decision") in 1884.
Impoundment	Accumulation or collection of water in a reservoir as a result of damming.
Inbreeding depression	A loss of vigor in a population associated with crosses between closely related individuals.
Instream use	A beneficial use of water in a stream channel as for recreation, fish and wildlife, navigation, the maintenance of riparian vegetation, or scientific study.
Jack	See *Grilse*.
Kesterson Wildlife Refuge	A 5,900-acre federal bird refuge and farm irrigation drainage reservoir in the western San Joaquin Valley. See *Selenium*.
King salmon	Common California name for chinook salmon.
Klamath Management Zone	The commercial ocean and river salmon fishing area where harvest of Klamath River salmon is regulated by the Pacific Fisheries Management Council.
Lateen rig	A sail device used on early Italian commercial salmon fishing boats in the San Francisco and Monterey bay areas. Lateen-rigged boats were adapted during the early twentieth century to accommodate internal combustion engines.
Levee	A natural or artificial ridge of material along a streambank that may contain or direct water flow.

Main stem	The principal channel of a river.
Mass wasting	A general term for a variety of processes by which large masses of earth material are moved by gravity either slowly or quickly from one place to another, as by landslides, slips, or creeps.
Mitigation	Avoiding, rectifying, or reducing impact; compensating for the impact by replacing or providing substitute resources or environments.
Mixed stock	Fish populations with members of varying origins—hatchery and naturally produced fish, for example, or fish produced in different stream systems.
Morphology	The study of the structure of organisms in terms of their form and the interrelationships of their internal parts.
National Environmental Policy Act (NEPA)	This 1969 law requires an environmental impact statement and consideration of alternatives before any major federal project may proceed. See *Environmental impact statement*.
Natural resource	Any portion of the natural environment, such as air, water, soil, forest, wildlife, and minerals.
Natural stock	Synonymous with native stock. Naturally produced fish populations evolved over aeons in a particular habitat.
Nursery area	Habitat, varying in nature from stream to stream, in which salmonids develop before smolting.
Offstream use	Use of water diverted from a naturally flowing stream.
Ozonization	The process of treating water with ozone, practiced in hatchery operations to control bacteria and viruses harmful to fish.
Pacific Coast Federation of Fishermen's Associations	A coalition of twenty-four commercial fishermen's groups noted for its activism at all levels of government to protect commercial fishermen's interests.

Peripheral Canal	A canal planned to bypass the Delta to deliver Sacramento River water to the Delta/Mendota Canal and California Aqueduct. Soundly defeated by referendum in 1982.
Poaching	Taking of fish or game by illegal means.
Production	The survival of fish to adulthood as measured by the abundance of recreational and commercial catch together with the return of fish to the state's spawning streams.
Public trust doctrine	The legal implication that natural resources are held in trust by the state and that the state as trustee has a duty and obligation to conserve and preserve them in the public interest. In California, the Audubon Decision was based on the public trust doctrine.
Race	A population or group of populations distinguishable from similar populations of the same species by genetic differences.
Rainbow	Rainbow trout; steelhead are anadromous rainbow trout.
Redd	Nest dug in stream gravels by female salmon and steelhead to deposit eggs.
Restocking	Replacing depleted fish stocks by providing new broodstocks of native or introduced populations.
Restoration	Fishery improvement activities—streambed and watershed improvement as well as maintenance of appropriate flow conditions—that seek to raise fish production and population numbers in a debilitated salmon or steelhead stream to the levels that obtained before a damaging incident or project.
Riparian	Related to the bank of a natural watercourse. Also refers to water rights of landowners whose property abuts a streambank.
Riprap	System of streambank protection often practiced by the Army Corps of Engineers involving em-

placement of concrete, rocks, or cobbles on basal slopes of streambanks.

Run A group of fish, usually migrating upstream as a unit.

Salmon Anadromous fish of the family Salmonidae. Six species of salmon have been reported in California, but by far the most common are chinook, coho, and steelhead.

Salmonids Fishes of the family Salmonidae. Three genera of salmonids are found in coastal California: the trouts (*Salmo*), chars (*Salvelinus*), and Pacific salmons (*Oncorhynchus*). In this book the term refers solely to California salmon and steelhead.

Saltery A commercial enterprise utilizing salt to preserve fish for storage or shipment. Canneries replaced salteries.

Sedimentation The settling of solids in any body of water because of gravity or chemical precipitation.

Seine net A mesh trap that encircles fish and draws them ashore or to a boat.

Selenium A naturally occurring element in soils that is essential to life but toxic in high concentrations. Kesterson Wildlife Refuge, in western San Joaquin Valley, is notable for its role in revealing the dangers to wildlife attempting to breed in waters contaminated with selenium.

Side-cast Excavated material from road construction or maintenance that is pushed over the side of the road.

Siltation Deposition of silt in streambeds that interferes with spawning and survival of eggs deposited in redds. Common causes of siltation are loss of stream basin vegetation, excessive erosion, and reduced streamflow.

Slough A slow-moving creek in a marshland or tidal flat or an inlet from a river.

Smolt	A juvenile seaward-bound salmonid in the process of transition in life cycle from fresh water to salt water.
Spawning	The reproductive act of male and female salmonids depositing and fertilizing eggs in a redd.
State Water Project	A statewide water development that is complementary and in some respects parallel to the Central Valley Project.
Steelhead	(*Oncorhynchus mykiss;* formerly *Salmo gairdneri*). Usually a sea-run rainbow trout. Sometimes called steelhead rainbow trout or steelhead salmon. Taxonomically a Pacific salmon.
Substrate	Streambed deposits that comprise a place where organisms grow or live.
Tagging	Systems of marking fish to provide information about populations, migratory and harvest patterns, and the like as a management tool.
Tractor yarding	Moving logs by tractor from where the tree was felled to a loading point for transport to a mill (as opposed to the use of steel cables and pulleys).
Transportation	The means by which a stream carries material: solution, suspension, and bed load.
Trawl net	A baglike net towed by a ship at specific depths for catching various forms of marine life.
Trolling	A means of fishing in which a lure or baited hook is pulled behind a moving boat. California commercial salmon fishermen may use only this method.
Turbidity	The degree of water's opaqueness due to the amount of fine matter in suspension. Salmon and steelhead reproduce best in water of low turbidity.
United States Government:	
Army Corps of Engineers	The corps manages and executes civil works programs, including a variety of activities related to

rivers, harbors, waterways, and administration of laws to protect and preserve navigable waterways. Like the Bureau of Reclamation, it is involved in federal water development projects in California, focusing on flood control and associated streambank protection.

Bureau of Indian Affairs (BIA)

An agency of the Department of Interior. The overall goal of the BIA is self-direction and management of programs for the benefit of Indians and Alaska Native people under a trust relationship with the federal government. It facilitates full development of their human and natural resources and trains them to manage their own affairs.

Bureau of Land Management (BLM)

An agency of the Department of Interior responsible for total management of 270 million acres of public lands not managed by other U.S. agencies (such as the Fish and Wildlife Service, Forest Service, Park Service), mainly in the West. This includes wild and scenic rivers and management of watersheds to protect soil and enhance water quality.

Bureau of Reclamation

The basic responsibility of the bureau, an agency of the Department of Interior, is to administer a reclamation program to provide secure water supplies for farms, towns, and industries in seventeen contiguous western states. It is responsible for river regulation, development of hydropower, outdoor recreation opportunities, and enhancement and protection of fish and wildlife habitats. In California it is the chief builder and operator of the Central Valley Project.

Council on Environmental Quality (CEQ)

An advisory body appointed by the president to formulate and recommend national policies to promote the improvement of environmental quality.

Environmental Protection Agency (EPA)

The purpose of the EPA is to protect and enhance our environment today and for future generations to the fullest extent possible under the laws enacted by Congress. Its mission is to work cooperatively with state and local governments to control and abate air and water pollution by solid waste, pesticides, radiation, and toxic substances.

Fish and Wildlife Service

An agency of the Department of Interior whose mission is to conserve, protect, and enhance fish and wildlife and their habitats for the continuing benefit of the American people. It assists, through a public information program and other means, in the development of an environmental stewardship ethic for our society based on ecological principles, scientific knowledge of wildlife, and a sense of moral responsibility. In California it assesses effects of federal water development projects on fish and wildlife resources and works with the Department of Fish and Game to develop mitigation programs.

Forest Service

An agency of the Department of Agriculture responsible for national leadership in forestry. Its mission is to provide a continuing flow of natural resource goods and services from National Forests to help meet national and international needs.

National Marine Fisheries Service (NMFS)

A unit of the Department of Commerce that conducts fishery research programs, administers the Marine Mammals Protection Act, and cooperates with coastal states concerning the protection of marine fish habitat, including that of the Pacific salmons. The NMFS provides technical services to regional fishery management councils established by the Magnuson Act.

Pacific Fisheries Management Council (PFMC)

A regional (Washington, Oregon, Idaho, California) agency of the Department of Commerce established under the National Fishery Conservation and Management Act to administer federal laws relating to control of fisheries from three miles to two hundred miles offshore.

Western Area Power Administation (WAPA)	An agency of the Department of Energy responsible for the federal marketing and transmission of electric power in fifteen central and western states.
Upmigrants	Salmon or steelhead moving into streams from the ocean or an estuary, usually to spawn.
Watershed	Originally a divide between drainage basins. Now often used synonomously with *drainage basins*.
Weir	Any structure across a watercourse that controls the movement of fish.
Wetland	Any area in which the water table stands near, at, or above the land surface for at least part of the year. Such areas are characterized by biota adapted to wet soil conditions.
Wild and Scenic Rivers System	A set of waterway protection laws enacted by the California legislature in 1972 affecting the Klamath, Smith, Trinity, and Eel rivers and part of the American River. These streams were later included under the corresponding national act.
Works Progress Administration (WPA)	A federally sponsored public works program to provide jobs for the unemployed during the Great Depression.

Suggested Reading

Books and Reports

Allen, Thomas B. *Guardian of the Wild: The Story of the National Wildlife Federation, 1936–1986.* Bloomington: University of Indiana Press, 1987.

Ashworth, William. *Nor Any Drop to Drink.* New York: Summit Books, 1982.

Bakker, Elna S. *An Island Called California.* 2nd ed. Berkeley: University of California Press, 1984.

Bearss, Edwin C. "History Resource Study, Hoopa-Yurok Fisheries Suit, Hoopa Valley Indian Reservation, Del Norte and Humboldt Counties, California." Washington, D.C.: USDI–BIA, 1981.

Berger, John J. *Restoring the Earth.* New York: Doubleday/Anchor, 1987.

Brown, Bruce. *Mountain in the Clouds: A Search for the Wild Salmon.* New York: Simon & Schuster, 1982.

California Advisory Committee on Salmon and Steelhead Trout. Reports to the Director, Department of Fish and Game. *An Environmental Tragedy* (1971); *A Conservation Opportunity* (1972); *The Time Is Now* (1975). Sacramento: California Department of Fish and Game.

———. Reports to the Joint Legislative Committee on Fisheries and Aquaculture and the Director, Department of Fish and Game. *The Tragedy Continues* (1986); *A New Partnership* (1987); *Restoring the Balance* (1988). Sacramento: California Legislature, Joint Publications Office.

California Department of Fish and Game. "Fish and Game Water Problems of the Upper San Joaquin River—Potential Values and Needs." Sacramento, 1951. Mimeo.

———. *Anadromous Fishes of California*, by Donald H. Fry, Jr. Sacramento: California Resources Agency, 1973.

California Division of Fish and Game. Fish Bulletin No. 17. *Sacramento–San Joaquin Salmon Fishery of California*. Sacramento: California State Printing Office, 1929.

———. Fish Bulletin No. 34. *Salmon of the Klamath River, California*, by John O. Snyder. Sacramento: Department of Natural Resources, 1931.

———. *Thirty-Eighth Biennial Report, 1942–1944*. Sacramento: California State Printing Office, 1944.

———. Fish Bulletin No. 98. *The Life Histories of the Steelhead Rainbow Trout and Silver Salmon*, by Leo Shapovalov and Alan C. Taft. Sacramento: Department of Fish and Game, 1954.

California Fish and Game Commission. *Twenty-Fourth Biennial Report, 1914–1916*. Sacramento: California State Printing Office, 1916.

California, State of. *Fish and Game Code of California*. New York: Gould Publications, 1989.

Dasmann, Raymond F. *The Destruction of California*. New York: Macmillan, 1965.

DeBell, Garrett, ed. *The New Environmental Handbook*. San Francisco: Friends of the Earth, 1980.

de Roos, Robert. *The Thirsty Land*. Stanford: Stanford University Press, 1948.

De Santis, Marie. *California Currents*. Novato, Calif.: Presidio Press, 1985.

Devall, Bill, and George Sessions. *Deep Ecology: Living As If Nature Mattered*. Salt Lake City: Peregrine Smith Books, 1985.

Everest, Fred H. *Western Anadromous Fish Habitat Program Plan, 1986–1990*. Portland: Pacific Northwest Forest and Range Experiment Station, USDA Forest Service, 1986.

Frome, Michael. *Conscience of a Conservationist*. Knoxville: University of Tennessee Press, 1989.

Gottlieb, Robert. *A Life of Its Own—The Politics and Power of Water*. San Diego: Harcourt Brace Jovanovich, 1988.

Hallock, Richard J. *Sacramento River System Salmon and Steelhead Problems and Enhancement Opportunities*. Sausalito: California Advisory Committee on Salmon and Steelhead Trout, 1987.

Kahrl, William L., ed. *The California Water Atlas*. Sacramento: State of California, 1979.

Kroeber, Theodora. *Ishi in Two Worlds*. Berkeley: University of California Press, 1961.

Leopold, A. Starker. *Wild California: Vanishing Lands, Vanishing Wild-*

life. Berkeley: University of California Press in cooperation with The Nature Conservancy, 1985.

London, Jack. *Tales of the Fish Patrol.* Oakland: Star Rover House, 1982.

McEvoy, Arthur F. *The Fisherman's Problem: Ecology and Law in the California Fisheries, 1850–1980.* New York: Cambridge University Press, 1986.

McGinnis, Samuel M. *Freshwater Fishes of California.* Berkeley: University of California Press, 1984.

McNeil, William J., ed. *Salmon Production, Management, and Allocation.* Corvallis: Oregon State University Press, 1988.

Margolin, Malcolm. *The Ohlone Way: Indian Life in the San Francisco–Monterey Bay Area.* Berkeley: Heyday Books, 1978.

Meyer Resources, Inc. *The Economic Value of Striped Bass, Chinook Salmon, and Steelhead Trout of the Sacramento–San Joaquin River Systems.* Sacramento: California Department of Fish and Game, 1985.

Moyle, Peter B. *Inland Fishes of California.* Berkeley: University of California Press, 1976.

National Geographic Society. *America's Wild and Scenic Rivers.* Washington, D.C.: Special Publications Division, 1981.

Nelson, Byron, Jr. *Our Home Forever: A Hupa Tribal History.* Hoopa, Calif. and Salt Lake City: University of Utah Printing Service, 1978.

Netboy, Anthony. *The Salmon: Their Fight for Survival.* Boston: Houghton Mifflin, 1974.

———. *Salmon: The World's Most Harassed Fish.* Tulsa: Winchester Press, 1980.

———. *The Columbia River Salmon and Steelhead Trout: Their Fight for Survival.* Seattle: University of Washington Press, 1980.

———. *The Memoirs of Anthony Netboy: A Writer's Life in the 20th Century.* Ashland, Oreg.: Tree Stump Press, 1990.

Nicholson, Max. *The New Environmental Age.* New York: Cambridge University Press, 1987.

Pacific Fisheries Management Council. *Review of the 1984 Ocean Salmon Fisheries.* Portland, 1985.

Palmer, Tim. *Stanislaus: The Struggle for a River.* Berkeley: University of California Press, 1982.

———. *Endangered Rivers and the Conservation Movement.* Berkeley: University of California Press, 1986.

Reichard, Nancy. *Stream Care Guide.* Eureka, Calif.: Redwood Community Action Agency, 1987.

Reisenbichler, R. R. *Report to USFWS on the Evaluation of Effects of Degraded Environment on Chinook Salmon in California Streams.* Red Bluff, Calif.: USFWS, 1986.

Reisner, Marc. *Cadillac Desert*. New York: Viking Press, 1986.

Ryman, Nils, and Fred Utter, eds. *Population Genetics and Fishery Management*. Seattle: University of Washington Press, 1987.

Salo, Ernest O., and Terrance W. Cundy, eds. *Streamside Management: Forestry and Fishery Interactions*. Proceedings of the symposium held February 12–14, 1986. Seattle: University of Washington, Institute of Forest Resources, 1986.

Schmookler, Andrew Bard. *The Parable of the Tribes: The Problem of Power in Social Evolution*. Berkeley: University of California Press, 1984.

Shepard, Paul, and Daniel McKinley, eds. *The Subversive Science: Essays Toward an Ecology of Man*. Boston: Houghton Mifflin, 1969.

Sheridan, David. *Desertification of the United States*. Washington, D.C.: Council on Environmental Quality, 1981.

United States Government. *Central Valley Basin*. 81st Cong., S. Doc. 113. Washington, D.C.: Department of Interior, Bureau of Reclamation, 1949.

——. *Central Valley Project Documents*. Part Two: *Operating Documents*. 85th Cong., H. Doc. 246. Washington, D.C.: U.S. Government Printing Office, 1957.

——. *United States Government Manual*. Washington, D.C.: Superintendent of Documents, 1989–1990.

U.S. Department of Interior. *An Investigation of Fish-Salvage Problems in Relation to Shasta Dam*. Special Scientific Report No. 10. Washington, D.C.: 1940.

——. *A History of River Basin Studies*, Vol. 1, by John L. Sypulski. Albuquerque, 1974. Mimeo.

——. Discussion Papers by Felix E. Smith. *Agricultural Wastewater and the Public Trust* (November 1986); *The Changing Face of California's Central Valley—Fish and Wildlife Resources and Bay/Delta Hearings* (November 1987); *The Friant Division and Water Marketing* (September 1988); *Water Development and Management in the Central Valley of California and the Public Trust* (June 1987); *Water: An Ecosystem Forever* (September 1986). Sacramento: U.S. Fish and Wildlife Service.

Vondracek, Bruce, and Robert Z. Callaham. *California's Salmon and Steelhead Trout—A Research and Extension Program*. Wildlands Resources Center Report No. 13. Berkeley: University of California, 1987.

Wallace, David Rains. *The Klamath Knot*. San Francisco: Sierra Club Books, 1983.

——. *Life in the Balance*. San Diego: Harcourt Brace Jovanovich, 1987.

Western, David, and Mary C. Pearl. *Conservation for the Twenty-First Century.* New York: Oxford University Press, 1989.

Articles

Baiocchi, Joel C. "Use It Or Lose it: California Fish and Game Code Section 5937 and Instream Fishery Resources." *University of California Law Review* (1980):431–460.

Bidwell, John. "Memoirs." *Century Magazine,* February 1891, p. 519.

Chapman, D. W. "Effects of Logging upon Fish Resources of the West Coast." *Journal of Forestry* 60(8)(1962):533–537.

DelVecchio Good, Mary-Jo. "Women and Fishing on the North Coast." *Ridge Review* 3(2)(1983):38–40.

Hoffman, Eric. "Saving the Steelhead." *Monterey Life,* March 1986, pp. 52–54.

Kirkpatrick, C. A. "Salmon Fishery on the Sacramento River." *Hutchings' California Magazine,* June 1860, pp. 530–534.

Kroeber, Alfred L., and S. A. Barrett. "Fishing Among the Indians of Northwestern California." *Anthropological Records* 21(1)(1960):1–210.

Leidy, R. A. "Distribution and Ecology of Stream Fishes in the San Francisco Bay Drainage." *Hilgardia* 52(8)(1984).

Pinchot, Gifford. "How Conservation Began in the United States." *Agricultural History* 11(October 1937):255–265.

Poole, William. "For the Sake of Salmon." *San Francisco Sunday Examiner and Chronicle,* June 28, 1987.

Sax, Joseph L. "The Public Trust Doctrine in Natural Resources Law: Effective Judicial Intervention." *Michigan Law Review* 68(1970):473.

Periodicals

Audubon. Magazine of the National Audubon Society. New York, bimonthly.

Bay on Trial. Newsletter of the Bay Institute of San Francisco. Sausalito, quarterly.

California Today. Newsletter of the Planning and Conservation League. Sacramento, bimonthly.

Creek Currents. Newsletter of the Urban Creeks Council. Berkeley, quarterly.

ECONEWS. Newsletter of the Northcoast Environmental Center. Arcata, monthly.

The Leader. Newsletter of the National Wildlife Federation. Washington, D.C., monthly.

Monterey Life. Monterey, monthly.

The Nature Conservancy Magazine. Arlington, Va., bimonthly.

Outdoor California. Publication of the California Department of Fish and Game. Sacramento, bimonthly.

Pacific Discovery. Publication of the California Academy of Sciences. San Francisco, quarterly.

pcffa FRIDAY. Newsletter of the Pacific Coast Federation of Fishermen's Associations. Sausalito, biweekly.

Ridge Review. Mendocino, quarterly.

Western Water. Newsletter of the Water Education Foundation. Sacramento, quarterly.

Videotapes

On the Edge. San Rafael: Coastal Resources Center.

Return of the King. Sausalito: Pacific Coast Federation of Fishermen's Associations.

Court Cases

Arnett v. Five Gill Nets. 48 Cal. App. 3d 454 (1975).

California Trout v. State Water Resources Control Board. 207 Cal. App. 3d 585 (1989).

Donnelly v. United States 228 U.S. 243 (1913).

Mattz v. Arnett. 412 U.S. 481 (1973).

National Audubon Society v. Superior Court. 33 Cal. 3d 419; 189 Cal Rpt 346, 658 P.2d 709 (February 1983).

Rank v. Krug. 90 Fed Supp 773 S.D. Cal (1950).

United States v. Forty-Eight Pounds of Rising Star Tea, etc. 35 Federal Reporter 403 (1888).

United States v. State Water Resources Control Board. 227 Cal Rpt 161 (1986).

Contributors

Stanley M. Barnes, consulting civil engineer of Visalia, California, represents users of State Water Project water and is chairman of the California Water Commission.

William T. Davoren is founder and executive director of The Bay Institute, Sausalito, California.

Mary-Jo DelVecchio Good is an associate professor of medical sociology at Harvard University Medical School; she conducted research on California's North Coast from 1981 to 1983.

Scott Downie, former teacher, is a commercial fisherman and chairman of the California Advisory Committee on Salmon and Steelhead Trout and the Commercial Salmon Trollers Advisory Committee. He received the 1988 Conservation Achievement Award from the California/Nevada chapter of the America Fisheries Society.

Richard J. Hallock, a retired California Department of Fish and Game marine biologist, is a recognized expert on Sacramento River salmon and steelhead.

Ken Hashagen, a senior fishery biologist, is coordinator of State Fish Hatcheries, California Department of Fish and Game, Sacramento.

Joel W. Hedgpeth, an expert on aquatic biology, is emeritus professor of oceanography, Oregon State University.

Patrick Higgins is a consulting fishery biologist and past president of the Humboldt chapter of the American Fisheries Society.

Eric Hoffman writes mostly about wildlife. Author of *Renegade Houses* and *Adventuring in Australia* and hundreds of stories published nationwide, he won the National Headliner and Maggie awards in 1980 for his contributions to *California Living.*

William K. Hooper, Jr., an associate governor of California Trout, Inc., is a managing partner of a hotel development and management company that operates five hotels in the Bay Area.

Richard Hubbard, retired assistant director, Pacific Southwest Forest and Range Experiment Station, most recently served as executive director of the California Natural Resources Federation, an affiliate of the National Wildlife Federation.

William M. Kier, with broad experience in state government, as well as natural resources management and policy, is consultant to the California Advisory Committee on Salmon and Steelhead Trout.

Alan Lufkin, retired educator, is active in several environmental groups. He is a member of the California Advisory Committee on Salmon and Steelhead Trout.

Bill Matson is a commercial fisherman who learned his craft by fishing with his father. He has also been a secondary school science teacher, counselor, and administrator in North Coast schools.

Paul McHugh is a feature writer for *Outdoors,* a section of the *San Francisco Chronicle.*

Philip A. Meyer is president of Meyer Resources, Inc., an economic consulting firm in Davis, California.

Ronnie Pierce is a marine biologist, consultant on Klamath River fisheries, and Indian historian.

William Poole writes his *Detours* column biweekly for *This World,* a section of the *San Francisco Sunday Examiner and Chronicle.*

Kim Price, Conservationist II with the California Conservation Corps, is work coordinator of the Salmon Restoration Project, Humboldt Center, Weott.

Nancy Reichard, who was born with a magnetic attraction to flowing fresh water, is director of Natural Resources Services for the Redwood Community Action Agency, Eureka, California.

Felix E. Smith is an environmental assessment specialist with the U.S. Fish and Wildlife Service, Sacramento.

William D. Sweeney is retired area manager for the U.S. Fish and Wildlife Service, Sacramento.

Jim Tarbell and Judy, his wife, publish the quarterly *Ridge Review* in Mendocino, California.

Dave Vogel is a fishery biologist and project leader of the U.S. Fish and Wildlife Service field office on the Sacramento River in Red Bluff, California.

George Warner is retired chief, Anadromous Fisheries Branch, California Department of Fish and Game.

Cindy Deacon Williams, a fishery biologist, is legislative representative for the Fisheries and Wildlife Division of the National Wildlife Federation, Washington, D.C. From 1985 to 1989, she was the Sacramento River winter chinook coordinator for the California/Nevada chapter of the American Fisheries Society.

Jack E. Williams is Fisheries Program manager for the Bureau of Land Management, Department of Interior, Washington, D.C. He was formerly with the Endangered Species Program of the U.S. Fish and Wildlife Service while stationed at the University of California, Davis.

Robert R. Ziemer is project leader and research hydrologist with the Pacific Southwest Forest and Range Experiment Station, USDA Forest Service, Arcata, California.

Index

Advisory committees: formation of, 22–23, 24, 30; reports of, 23–26, 30–31. *See also* California Advisory Committee on Salmon and Steelhead Trout; Conservation, advisory committee reports

"Agribusiness," origin of term, 58

Alameda Steelhead Restoration Committee, 255–56

American Fish Culturists Association, 13, 53

American Fisheries Society, 186n; address by Livingston Stone to, 52; and Sacramento River winter chinook listing, 25, 105–15

Anderson-Cottonwood Irrigation District, 110, 228, 229

Artificial propagation. *See* Fish hatcheries

Asian driftnet fishery, 35–36

Atlantic salmon fishery, 8

Audubon Decision. See *California Trout v. State Water Resources Control Board; National Audubon Society v. Superior Court*

Baird, Spencer Fullerton, 53, 54

Baird Hatchery, 13, 53–54, 55, 58, 81, 107

Baumhoff, John, 42–43

Bay-Delta hearings, 195, 207–8. *See also* State Water Resources Control Board

Bay Institute of San Francisco, 28

Behr, Peter, 179–85

Bodega Bay, annual salmon festival, 8

Bondoux, Susan, 141

Bosco, Doug, 184–85

Boswell Company, J. G., 218

Boucherg, Jerry, 183

Bradley, William, 60

Brown, Edmund G. ("Pat"): as attorney general, 18, 62; as governor, 19–20

Bureau of Indian Affairs, 144, 149–50

Bureau of Land Management, 95, 181, 185

Bureau of Reclamation: and Coordinated Operation Agreement, 208; and CVP dams, 22; and CVP development, 15–17; fisheries policies of, 15–17, 27, 68–69; and Friant contract renewals, 34–35; and Kesterson Reservoir, 206; and Mono Lake case, 211; and political alliance, 210; and Red Bluff Diversion Dam and Trinity River, problems at, 217; and Red Bluff Diversion Dam, operation of, 96–98; and Red Bluff Diversion Dam, protection of fish at, 130–33; and Sacramento River winter chinook restoration plan, 111, 112, 115; and San Joaquin River, control of, 62, 65, 67; and Shasta Salvage Plan, 15–17; and Shasta water temperature controls, 32, 112–13. *See also* Central Valley Project; Public trust doctrine; Red Bluff Diversion Dam; Water development

Burt, Barry, 256

Bush, Jayne, 140–41

Caddis Flyer, 61n

California Advisory Committee on Salmon and Steelhead Trout: and classroom education project, 256; organization of, 30; reports of, 30–31, 95, 236; statewide goal of, 263; and Wildland Resources Center workshop, 89. *See also* Advisory committees; Conservation

146; and fisheries agreement with state, 28; and Fisheries Conservation and Management Act, 26; and Fisheries Management Council plan, 145; and gillnetting on Klamath River, 85, 143–45; and harvest rate management plan, 145–47; and Klamath canneries closure, 26, 143; and Klamath contribution rate, 32, 146–47; and lower Klamath fishery, 142–50; and relations between Indians and commercial trollers, 26, 27, 142–43. *See also* Commercial salmon fishing; Klamath River salmon management

Indian fishing rights: issues related to, 41–45; legality recognized, 7, 26, 28, 144; protections by Hupa Tribe, 142n

Indian reservations: confusion over boundaries of, 42, 43–44; fishing regulations on, 144–45; and Four Reservations Act of 1864, 42; homesteading on Hoopa Valley reservation, 42–44; and Klamath basin tribal goals, 26, 149–50; lower Klamath River, 41–45; and "navigable water" question, 43–45; and protection by U.S. Army, 42; and Yurok Reservation, 150

Indians, aboriginal: conservation practices of, 6, 39–40; customs and rituals of, 6–7, 38–40; employed by non-Indians, 7, 8, 42–43, 44, 142–43; fishery, early history of, 37–45; fishing methods of, 6, 39–40, 54; and impact of gold mining on, 41, 42; and interdependence of Hupa, Karok, and Yurok on fish, 38; and treaties, 41; white settlers' attitude toward, 41; and Yurok tribe, in pre-contact era, 37–40

Iron Gate Hatchery, 82, 232
Iron Mountain Mine, 110, 111

Johnson, Huey, 184
Jones, Martin Van Buren, 42

Karlton, Lawrence, 35
Keene, Barry: as assemblyman, 231; as senator, 234
Kelsay, Bill, 260
Kennedy, David, 60, 186
Kesterson Reservoir, 206, 218
Kingfisher Flat Steelhead Fish Hatchery, 258, 259, 261
Klamath River Anglers Association, 142
Klamath River salmon management: and Fisheries Resource Plan, 149; and harvest rate management plan, 145–47; and Klamath contribution rate, 32, 146–47; and Klamath Management Zone plans, 32–33, 145–47; and Klamath River Salmon Management Group, 145; problems of, 32–33, 155, 157–58
Klamath-Trinity: fisheries restoration project, 28, hatchbox-pond rearing project, 233
Krug, Julius, 15–16

Lake Red Bluff Power Project, 103–4
Landers, Keith, 245
Law enforcement: Fish and Game Code Section 5937, 14, 25, 200; Indian fishery protection laws, 7–8; local resentment of, 13; patrol boat on Sacramento River, 14
Leopold, Luna, 159–60
Limited entry program, 28, 230–31
Little River, Salmon Stamp Project, 234
Livermore, Ike, 179
Logging. *See* Forestry and fisheries
Los Angeles Department of Water and Power, 211, 219
Los Angeles Star, 41
Los Banos salmon trap, 235
Low-input sustainable agriculture, 265
Lujan, Manuel, 35, 218

McCarthy, Leo, 183–84
McCaslin, Matt, 256
McCrary, Lud, 259–60
McEvoy, Arthur F.: and *The Fisherman's Problem*, 8; on Indian conservation practices, 40
McKee, Redick, 41
Mad River Fish Hatchery, 193, 235
Magnuson Act, 27
Man and Nature (Marsh), 163
Marin Municipal Water District, 256
Marin Rod and Gun Club, 102
Marine Mammals Protection Act, 28
Marsh, George Perkins, 163
Marshall Plan, 15
Mattole River: Restoration Council, 248; Salmon Stamp Project, 234; Salmon Support Group, 249-50
Mattz v. Arnett, 144
Mendota Dam, 61–62, 63, 65
Metropolitan Water District of Southern California, 183, 218, 219
Michigan Law Review, 212
Migratory Bird Treaty Act, 213–14
Miller, Lyndsey, 140

Wildland Resources Center, 94
Wildlife, Coastal, and Park Land Conservation Act, 28
Wilson, Richard, 179
Women and fishing: changing role of, 141; and cultural barriers, 135–38; and man-woman partnerships, 138–40; on North

Coast, 135–41; and Noyo Women for Fisheries, 135; and Pacific Coast Fishermen's Wives Coalition, 135; and Salmon Restoration Association, 135
Works Progress Administration (WPA), 58

Zilly, Thomas S., 112

Compositor: Huron Valley Graphics
Text: 11/13 Caledonia
Display: Caledonia